Matthew Moncrieff Pattison Muir

The Story of the Wanderings of Atoms

Especially those of Carbon

Matthew Moncrieff Pattison Muir

The Story of the Wanderings of Atoms
Especially those of Carbon

ISBN/EAN: 9783337194888

Printed in Europe, USA, Canada, Australia, Japan

Cover: Foto ©berggeist007 / pixelio.de

More available books at **www.hansebooks.com**

THE STORY OF
THE WANDERINGS OF ATOMS

PREFACE.

IN attempting, in *The Story of the Chemical Elements*, to set forth some of the guiding conceptions of chemistry, I gave a sketch of the theory of the grained structure of matter. That theory has put an instrument into the hands of the chemist which has proved of signal service in setting in order the connexions between the compositions and the properties of bodies, and in making possible and developing new chemical industries.

To grasp the conceptions of the atom and the molecule, and to be able to apply these in detail, is demanded alike of the man who wishes to follow the development of chemical science, and of him who desires to be a successful manufacturer in any of the more recently established branches of the chemical trades. And, to follow some of those investigations which deal with the notions of molecular structure and molecular symmetry is an excellent exercise in accurate and imaginative reasoning, and in the use of theories and hypotheses as instruments for gaining knowledge.

As the ideas of the atom and the molecule have been more fully applied in the elucidation of the changes undergone by compounds of carbon than elsewhere in chemistry, I have confined myself in this volume to the compounds of that element.

M. M. PATTISON MUIR.

CAMBRIDGE, *August* 1899.

CONTENTS.

THE STORY OF

THE WANDERINGS OF ATOMS;

ESPECIALLY THOSE OF CARBON.

CHAPTER I.

INTRODUCTION.

IN *The Story of the Chemical Elements* mention
was made of certain changes in the combinations and collocations of elementary substances
which are constantly occurring in plants and
animals. In this volume I shall endeavour to
tell portions of the chemical story of these transmutations, and of transmutations like these.
The ripening of fruit is accompanied by the
change of starch into sugar, the disappearance
of acidic substances, the production of coloured
bodies which give the bloom to the ripening fruit,
and the formation of compounds which impart to
the ripened fruit its peculiar and agreeable aroma
and flavour. During the growth of certain kinds
of trees and shrubs there are produced various
gums, resins, and balsams; substances that are
valuable for their healing properties, or are
sought after because of the pleasant odours they
emit either in their natural state or when they
are burned. The products of the life-processes of
other plants are prized as medicinal agents; such
are quinine, belladonna, and oil of wintergreen.

There are plants and trees whose leaves, or fruit, contain compounds that are used as beverages, or to afford solace to human beings: everyone drinks a decoction of the compounds formed in the leaves of the tea-plant, or in the seeds of the coffee-tree; or forgets some of his cares while inhaling the smoke of tobacco. From the compounds produced in growing plants are formed other substances which find manifold uses; in this class may be named alcohol, chloroform, iodoform, ink, indigo, and paper.

As examples of large industries that depend upon the transformations of substances of vegetable origin into materials useful to mankind, it will suffice to mention the production of lubricating and burning oils, and paraffin-wax, the manufacture and the dyeing of cotton fabrics, the making of aniline and alizarin colours, and such processes as baking and biscuit-making.

A great part of the art of cooking consists in bringing about changes in the combinations and arrangements of the elementary substances of which the flesh of animals is composed. If living animals were not transformers of vegetable compounds into other combinations of the elements of these compounds, such industries as making butter, cheese, leather, lanoline, Prussian blues, and silk-stuffs, would have no existence. And, think of the strange transformations which occur when the food we consume is changed into the substance of our bodies, and our very bodies serve as material from which new tissues are constructed. The human being is constantly feeding on himself. As Sir Thomas Browne puts it

—" We are what we all abhor, *anthropophagi*, and cannibals, devourers not only of men, but of ourselves ; and that not in an allegory but a positive truth ; for all this mass of flesh which we behold came in at our mouths ; this frame we look upon hath been upon our trenchers ; in brief, we have devoured ourselves."

Agriculture is very closely connected with the selection and application of those kinds of food which certain classes of plants are able to assimilate rapidly, and to transmute into compounds which, in their turn, form the basis of foods for animals and human beings.

There are departments of preventive medicine which primarily depend on the formation of mixtures of compounds wherein certain minute organisms flourish, which organisms when injected into a human being, or an animal, cause changes such that the living being becomes immune, or partially immune, to the attacks of certain diseases. When an animal, or a plant, dies, changes in the arrangements and combinations of the elements continue ; many of the compounds that are formed are harmful to living human beings, and some of these compounds are deadly poisons. But there are minute living organisms that feed upon these harmful, or deadly compounds, and produce other collocations of elements which are devoid of ill effects on human beings. For instance, there are bacteria which feed on the compounds contained in sewage, destroying thereby the obnoxious substances in the sewage, and producing compounds which are perfectly harmless.

And finally, the chemist in his laboratory brings about rearrangements of the elements whereof compounds formed in living organisms are composed, and thus produces a great array of new bodies, many of which find important applications in manufactures or in medicine.

In this volume I shall attempt to draw the outlines of the chemical story of some of the compounds formed in living organisms, and of other compounds allied to these.

The object of chemistry is to elucidate the changes of composition which compounds exhibit, and to connect these with changes of properties.   Many illustrations of the ways whereby chemistry conducts the study of changes of composition and properties have been given in *The Story of the Chemical Elements*.   It is the business of this volume to apply these methods to a particular class of compounds, and to indicate some of the general results that have been obtained.

A brief statement of the more general results that were gained by an examination of the compositions and the properties of definite kinds of matter in the former volume will be given here; but it will be assumed that the reader has made himself acquainted with these matters in some detail.   A compound is a definite kind of matter which cannot be separated into portions unlike one another, and unlike the original substance, by any process of the nature of sifting or sorting. The composition of a compound is stated in chemistry by naming the elements into which the compound can be resolved and by the union of which the compound can be produced, and by indicating the quantity by weight of each element that can be obtained from, or is used in the formation of, some determinate weight of the compound. So far as our knowledge goes at present, an element is a substance all the particles whereof, however small these particles may be, are alike in every respect; except, of course, that one particle may be heavier, or lighter, than another.

The properties of the product of the combination of two or more elements, which product is called a compound, are unlike, and generally very unlike, the properties of the elements by whose union the compound is formed : but it is always possible to tear the compound asunder, and to obtain again the elements that seemed to lose their identity in the act of producing the compound. The information that is gained by the study of the compositions of compounds is expressed by the chemical formulæ which are assigned to compounds. These formulæ are expressions of the compositions of compounds in the terms of a special language. The formula of a specified compound tells the number of combining weights of each of the elements which have combined to form a reacting weight of that compound. It is possible to express the composition of a compound by a formula because it is always true, and always exactly true, that *all elements react in the ratios of their combining weights, or of whole multiples of their combining weights.*

The study of the properties of elements and compounds is more complex than the study of

composition; and the results of the investigations concerning properties cannot be stated in
such generalised and yet perfectly exact terms as
those which express the results of the study
of composition. The chemical properties of an
element, or a compound, are those which the
body exhibits when it interacts with other
elements and compounds; these properties are,
therefore, better named *reactions*. The bodies
which take part in a chemical reaction are
changed into bodies unlike themselves. For
instance—any specified weight of the element
oxygen combines with one-eighth of its own
weight of the element hydrogen, the product is
water, and the weight of the product is exactly
the sum of the weights of the two elements
which have passed out of existence, as such, in
the act of producing this quantity of water;
these facts may be grouped together and called
a reaction of hydrogen, or they may be called a
reaction of oxygen, or they may be called a reaction of water.

The study of the reactions of elements and
compounds, taken with the study of composition,
has enabled us to use such generalised terms as
acids, alkalis, salts, and the like. All compounds
which are called acids have one common reaction,
and one thing in common as regards their composition : they are all compounds of hydrogen,
and, in the presence of water, they all interact
with iron, lead, copper, zinc, or some other
element which is a *metal*, to exchange the whole
or a part of their hydrogen for metal. The
reactions of compounds are conditioned by the

kinds of elements of which they are composed.
But there exist compounds of the same elements
which are wholly unlike one another in their
reactions: for instance, the compound whose
composition is expressed by the formula NH_3 is
an alkali; but the compound of nitrogen and
hydrogen whose composition is expressed by the
formula N_3H is an acid. It is evident that the
reactions of compounds are conditioned not only
by the kinds of elements, but also by the relative
quantities of the elements, of which the compounds consist.

Many of the differences that are observed
between the reactions of compounds are pictured
to himself by the chemist as accompaniments of
differences in the kinds, and differences in the
quantities, of the elements which compose these
compounds. But all differences in the reactions
of compounds cannot be connected with differences of composition, unless the word *composition*
is used with a more extended signification. There
are compounds which shew very different reactions although equal weights of these compounds are composed of the same quantities by
weight of the same elements. The case of the
two compounds urea and ammonium cyanate,
the composition of both of which is expressed
by the formula N_2CH_4O, was considered in
Chapter IX. of *The Story of the Chemical Elements*.
One hundred parts by weight of urea are composed
of $46\frac{2}{3}$ parts by weight of nitrogen, $26\frac{2}{3}$ parts by
weight of oxygen, 20 parts by weight of carbon,
and $6\frac{2}{3}$ parts by weight of hydrogen; and one hundred parts by weight of ammonium cyanate have

also exactly this composition. But the reactions of these compounds are very different. In certain cases, then, differences in the reactions of compounds must be conditioned by some circumstance besides differences in the kinds and the quantities of the constituent elements. The only other conditioning circumstance that can be thought of clearly is, differences in the arrangements of the constituent elements.

The moment an attempt is made to apply the conception of a definite arrangement of the elements that form a compound so as to frame a clear and consistent mental picture of the connexions that are assumed to exist between this arrangement and the reactions of the compound, it is discovered that some theory of the structure of matter must be adopted. The theory that is always employed in chemistry is that which likens the structure of matter to the structure of a barrelful of apples or oranges, or to that of a brick wall, or of a quantity of small shot. The extremely minute particles whereof a quantity of any definite kind of matter is composed, according to this theory, are called *molecules*. The properties of any element or compound are asserted by the theory to be the properties of the molecules of that body. Now, the fact that a compound can be broken up into simpler compounds and then into certain elements, is explicable in terms of the molecular theory only by assuming the existence of particles of matter weighing less than the molecule. The parts of molecules are called *atoms*. The theory presents the molecules of a body to our mental vision as

collocations of atoms ; and it enables us to think
clearly of the properties of these molecules as
conditioned by three circumstances, firstly the
kinds of atoms that compose the molecules,
secondly the number of atoms of each kind, and
thirdly the arrangement of the atoms : or, as
Lucretius said, 1900 years ago, "it matters much
with what others, and in what positions, the same
first-beginnings of things are held in union, and
what motions they do mutually impart and re-
ceive. The same numbers of the same atoms
may be arranged in different ways ; the results
of the different arrangements will be molecules
whose reactions are not the same." *

The application of the molecular and atomic
theory to the problem of representing the
arrangement of the parts of molecules in such a
way as shall make it possible to connect chemical facts, clearly, consistently, and suggestively,
with the theoretical presentment of these facts,
necessitates the use of certain subsidiary hypotheses and certain conventions. The most important hypothesis is that which asserts the
atom-fixing power of each atom in a molecule to
be limited and definite. Facts like those referred to on p. 169, and the following pages, of
The Story of the Chemical Elements, can be interpreted in terms of the molecular and atomic
theory only by saying that, "there is a limit to
the number of atoms of any kind wherewith a
specified atom can enter into direct chemical

* For a more detailed discussion of the molecular and
atomic theory, the reader is referred to Chapter VIII. of
The Story of the Chemical Elements.

B

union to produce a molecular building which does not fall to pieces." The hypothesis that each atom in a molecule is able to link directly to itself a certain limited number of other atoms is forced on us by such facts as these: the existence of the molecule NH_3 and the impossibility of forming a molecule composed of more than three atoms of hydrogen and one atom of nitrogen, the existence of the molecule H_2O and the impossibility of forming a molecule composed of more than two atoms of hydrogen and one atom of oxygen, the existence of the molecule CH_4 and the impossibility of forming a molecule composed of more than four atoms of hydrogen and one atom of carbon, and the existence of the molecule HCl and the impossibility of forming a molecule composed of more than one atom of hydrogen and one atom of chlorine. It is customary to say that the atom-fixing power of the carbon atom is twice that of the oxygen atom, and is four times greater than the atom-fixing power of the atom of chlorine. The expression *atom-fixing power of a specified atom* means the maximum number of atoms between which and the specified atom direct action and reaction occurs in any molecule: the atom-fixing power of an atom is measured by the maximum number of atoms of hydrogen with which the specified atom combines to form a molecule. The term *valency* of an atom is generally employed as synonymous with atom-fixing power. The conventions adopted for expressing the valencies of atoms are two; the symbol of the element is written with a Roman numeral above it, or straight lines equal in number to the valency of the atom are attached to the symbol of the element.

The molecule of a compound is thought of in chemistry as a definite structure; each atom is pictured as in direct connexion with a limited number of other atoms; the reactions of the molecule are regarded as dependent on the kinds of atoms, the number of each kind of atoms, and the way in which the atoms are linked together. This is the only hypothesis which has been found capable of bringing order into the enormous array of chemical facts concerning the compositions and reactions of compounds.

Those formulæ of compounds which represent the structure or constitution of the molecules of these compounds in the language of the hypothesis of atom-linking are called *structural* or *constitutional*, or sometimes *rational, formulæ.* For instance, two compounds are known to be formed by the union of four atoms of hydrogen with two atoms of carbon and two atoms of chlorine; the formula $C_2H_4Cl_2$ expresses the composition of both compounds. To one compound is assigned the constitutional formula

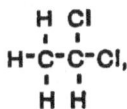

H Cl
ı ı and to the other the   H H
H-C-C-Cl, constitutional formula Cl-C-C-Cl.
ı ı   ı ı
H H   H H

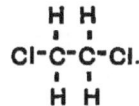

In one molecule, both atoms of chlorine are represented as in direct union with the same atom of carbon. In the other molecule, the chlorine atoms are represented as directly linked to different atoms of carbon. These compounds are

formed by different processes, and the reactions of
the two compounds differ ; the formulæ are expressions of the reactions of formation, and the
reactions of decomposition, of the compounds
in a special and extremely symbolic language,
which has gradually grown as the necessity
has been felt for a suitable means of presenting
the facts of chemical composition and chemical
change.   Each formula is a sentence in this
language ; it is not a realistic presentment of
the molecule itself.   If we are to think clearly
about the properties of matter we must think
of matter as having a grained structure ; but
the conception the chemist has formed of a
molecule may be, nay almost certainly is, extremely unlike the actual structure of the minute
particles of any kind of matter.   The very term
matter is only a convenient symbol.   We must
bring together, compare, contrast, generalise the
facts regarding composition and reactions.   The
first necessity is to express the facts in a clear,
descriptive, and suggestive language.   The
chemical past is strewn with the fragments of
dead languages.   The only mode of expression
which has been found capable of expansion and
modification is that which assumes the molecule
to be a definite structure of atoms, and enables
us to think clearly about this structure by
picturing it to ourselves as held together by
direct actions and reactions between the individual atoms.   As we proceed with the special
portion of the chemical story which concerns us
in this volume, we shall be convinced of the impossibility of making progress without a suitable

vehicle of expression ; and we shall recognise both the elasticity and the imperfections of that medium by means of which the facts of chemical composition and change are presented.

CHAPTER II.

A SURVEY OF THE COMPOSITION AND REACTIONS OF THE COMPOUNDS OF CARBON.

IF we consider the enormous number of living things in the world, and, withdrawing our thoughts from every other aspect of these living organisms, we regard them only as active producers of chemical compounds, we shall be over-whelmed with the vastness of the problem that is presented to the chemist. For a part of the chemist's calling is to examine the changes in the collocations of elements that are proceeding every moment in living things, or by the agency of living things, to sort and classify the compounds that are produced, to compare and contrast the processes of change, and to endeavour to express the knowledge thus accumulated in the fewest possible general statements. Moreover the examination of the compounds formed in organised beings, or by their aid, has led to the discovery of a more numerous class of compounds similar to these, although produced without the help of living laboratories. These compounds also must be analysed and synthesised ; their reactions must be discovered and chronicled ; and the resemblances and differences between them must be elucidated. And then, the sum of knowledge expressed by the words *organic chemistry* must be brought into vital connexion with the other parts of the science of chemistry; and the threads that bind this portion of our knowledge of nature with the rest of the domain of natural science must be disentangled, and made ready for picking up at the time when all parts of natural knowledge shall be seen to be members of one body " fitly joined together."

The first thing to be done in the chemical examination of compounds is to find their compositions. An examination of the compositions of the compounds produced in living organisms, or by the help of these organisms, shows that every one of them is a compound of the element carbon. And those compounds which are allied to the products of organisms, but are produced in the laboratory, are found also to be compounds of carbon. The elements with which carbon is combined in by far the greater number of the compounds referred to are these three—hydrogen, oxygen, and nitrogen.

Carbon is an element which exists in more than one form. Diamond is almost pure carbon; graphite is a mixture of carbon with other substances, and it is possible to remove these substances and obtain pure graphite; pure amorphous carbon is formed by calcining cane sugar in a covered vessel and treating the black residue with various chemical reagents. If equal weights of diamond, pure graphite, and pure amorphous carbon are burnt in plenty of oxygen, the only product is carbonic acid gas, and the same weight of this gas is obtained in each case. The compounds formed by combining carbon with other elements are the same whether the carbon is used as diamond, graphite, or pure charcoal. These facts oblige us to say that diamond, pure graphite, and pure amorphous carbon are forms or varieties of the same element. When diamond is heated to an exceedingly high temperature, out of contact with air or oxygen, it glows, swells, and splits, and after cooling the surface resembles graphite. Amorphous carbon has been changed to the diamond form of the element by dissolving it in cast-iron, melting the carburetted iron in an electric furnace, and allowing it to fall into mercury covered with a layer of water; the spheres of iron which form are found to contain very minute transparent diamonds.

There are other elements which exist in different forms; for instance, arsenic, phosphorus, and oxygen. The differences between the two varieties of arsenic are not very marked; one form is crystalline, the other is amorphous (that is, without crystalline structure): amorphous arsenic changes into crystalline arsenic when it is heated for some time to a low red heat, out of contact with air. Ordinary, or yellow, phosphorus differs very much from red, or amorphous, phosphorus. The yellow variety is a soft, semi-transparent, colourless, crystalline, solid, with a distinct smell; it glows in the dark, smokes when exposed to the air, and takes fire very easily; it melts at 44° C. [111° F.], dissolves readily in bisulphide of carbon, and is extremely poisonous. Red phosphorus is a dark

carmine-coloured, opaque, powder without taste
or smell; it does not glow in the dark, and it
must be heated to about 250° C. [482° F.] before
it takes fire; it does not melt at a red heat, nor
does it dissolve in bisulphide of carbon, and it is
not poisonous. The differences between these two
varieties of phosphorus are so great that one would
seem to be justified in saying the substances are
different kinds of matter. Nevertheless the products of the union of phosphorus with other
elements are the same whether these compounds
are formed from the yellow, or from the red,
variety: in other words, there is only one class
of compounds of phosphorus, and it is a matter
of complete indifference whether the compounds
are produced from yellow phosphorus, or from
red phosphorus, as a starting point. Moreover
equal weights of the two kinds of phosphorus
are changed into the same quantities by weight
of the same compounds; for instance, if one grain
of either variety is burnt in plenty of oxygen,
2·29 grains of an amorphous, snow-like solid—
called phosphorus pentoxide—are produced. (The
change begins at the ordinary temperature of the
air when yellow phosphorus is used, but not till
a temperature approaching 500° F. is reached
when red phosphorus is employed.) Again if
one pound of yellow phosphorus is warmed in a
stream of chlorine, there are produced 4·435 lbs.
of a clear, colourless, liquid, which fumes in the
air and gives off a vapour that attacks the eyes
in a most disagreeable way; and the same weight
of the same liquid—called phosphorus trichloride
—is obtained by warming one pound of red phos-

phorus in chlorine. Finally, yellow phosphorus
is changed into red phosphorus by heating to
about 250° C. [482° F.] in a vessel filled with
carbonic acid gas, or nitrogen, or some other
gas which is without chemical action on phosphorus; and the facts that red phosphorus
is the only product, and the weight of red
phosphorus formed is the same as the weight
of yellow phosphorus used, shew that nothing
has been added to, and that nothing has been
taken away from, the phosphorus in the process. The reverse change, that is to say, the
change of red into yellow phosphorus, is effected
by heating the red variety to about 300° C.
[572° F.] in a vessel filled with an indifferent
gas; the weight of yellow phosphorus obtained
is the same as the weight of red phosphorus
used.

The manufacture of *safety-matches* depends on
the facts that red phosphorus is not poisonous;
that it can be produced by heating ordinary
phosphorus in a gas which does not react with
the phosphorus; and that it cannot be ignited
by rubbing, unless the surface on which it is
rubbed contains substances rich in oxygen and
ready to give up oxygen to the phosphorus. The
heads of safety-matches are tipped with a mixture
of such compounds as chlorate of potash, chrom-
ate of potash, peroxide of lead, and sulphide of
lead; and the rubbing surface on the box is a
mixture of red phosphorus and powdered glass,
sometimes containing antimony sulphide or man-
ganese peroxide, made into a paste with glue.
When the match-head is rubbed on the prepared

surface the oxygen-containing compounds are partially decomposed, oxygen is produced and combines with some of the red phosphorus on the rubbing-surface; the act of combination produces sufficient heat to cause the ignition of the compounds of sulphur in the match-head, and the flame is then passed on to the wooden stem of the match.

The element oxygen exists in two modifications; ordinary oxygen, and ozone; both are gases at the ordinary temperature of the air. When oxygen is submitted to the action of the silent electric discharge a portion of the oxygen is changed into a substance which is a more energetic oxidiser than oxygen. This substance converts mercury, at the ordinary temperature, into black oxide of mercury, a reaction which is not accomplished by oxygen; when brought into contact with a solution of iodide of potassium it produces oxide of potassium, iodine, and oxygen, this also being a reaction which oxygen does not effect. When oxygen which has been submitted to the action of the silent electric discharge is cooled to about *minus* 180° C. [*minus* 292° F.] a blue liquid is formed; if this is allowed to evaporate oxygen passes off as a gas, and the modification of oxygen called ozone remains, presenting the appearance of a very dark blue liquid, and boiling at *minus* 106° C. [*minus* 214° F.]. Liquid oxygen boils at about *minus* 181° C. [*minus* 294° F.]. Ozone is completely changed into oxygen by heating to low redness; the weight of the oxygen obtained is equal to the weight of the ozone used; but the volume of the oxygen is one and a half times as great as the volume of the ozone. Ozone is one and a half times heavier than oxygen, bulk for bulk ; and, as a contraction of volume attends the formation of ozone from oxygen, ozone may be called condensed oxygen. As both oxygen and ozone are gases under the ordinary conditions of temperature and pressure, it is possible to find their molecular weights (see *The Story of the Chemical Elements*, pp. 156 *and following*). The result is that while a molecule of oxygen is 16 times heavier than a molecule of hydrogen, a molecule of ozone is 24 times heavier than a molecule of the standard element, hydrogen. Now, as the molecular weight of hydrogen is represented by the number *two* (see *Chemical Elements*, pp. 160-161), it follows that the molecular weight of oxygen is 32, and the molecular weight of ozone is 48. The atomic weight of oxygen is 16 (*Chemical Elements*, pp. 161-163). The molecules of oxygen and ozone are composed of the same kind of atoms, namely, atoms of oxygen ; the molecular symbol of the former is O_2, and the molecular symbol of the latter is O_3.

The consideration of oxygen and ozone, then, shows that some of the properties of a molecule composed of two atoms of the same kind may be different from the properties of a molecule composed of three atoms of the same kind as those which formed what we may call the diatomic molecule. This is a particular instance of the phenomenon already noticed, that the reactions of a molecule are conditioned, among other circumstances, by the number of the atoms which compose it. The present instance is especially interesting, because the atoms which compose the molecules of oxygen and ozone are all of the same kind.

The molecular weights of red and yellow phosphorus are not known with certainty, nor are the molecular weights of the three varieties of carbon known ; but, arguing from analogy, on the basis of the similar phenomena presented by oxygen and ozone, it is probable that the number of atoms which compose the molecule of one or the modifications of phosphorus is different from the number of atoms which compose the molecule of the other modification of that element ; and that the molecular weights of the three kinds of carbon are different.

Carbon is rather an inert element ; it does not very readily enter into union with other elements ; for instance, combination with oxygen begins only at about a red heat. Carbon belongs to the class of non-metallic elements ; it is altogether different in its chemical functions from such elements as iron, copper, zinc, and lead. We have then to consider the compositions and the reactions of the compounds which are formed by the union of this element carbon with other elements, and more particularly with the three elements hydrogen, oxygen, and nitrogen. Hydrogen is the lightest known form of matter ; it is a colourless gas, without smell or taste. Hydrogen burns when a lighted match is brought near it, provided air, or oxygen, is in contact with the hydrogen ; the product of the burning is water. The part played by hydrogen in compounds is conditioned by the nature, and the quantities, of the elements with which it is combined. For instance, compounds of hydrogen with comparatively large quantities of non-metallic elements, such as sulphur, chlorine, and oxygen, are acids; whereas the compounds of hydrogen with relatively large quantities of the most markedly metallic elements, in addition to oxygen, are alkalis. (Compare *Chemical Elements*, pp. 81 to 90.) The elements oxygen and nitrogen are colourless, odourless, tasteless, gases. The former very readily combines with most of the other elements to form oxides; the latter belongs to the class of inert substances. The compounds which oxygen forms differ exceedingly in their chemical reactions; some are acidic, others are basic, and others are salts. Most of the acids it is true are compounds of oxygen, but in the class of compounds of oxygen it is necessary to place the alkalis also. (Compare *Chemical Elements*, pp. 109 to 113.) Nitrogen forms a compound with hydrogen which is a marked alkali; and it also forms a compound with the same element which has the characteristic reactions of an acid. Oxygen is pre-eminently *the* supporter of combustion; in the burning of coal, wood, coke, oil, wax, and other sorts of fuel, it is the oxygen in the air which combines with the elements, especially with the carbon of the fuel, and produces the heat and light that we feel and see. If the atmosphere contained a relatively very large quantity of oxygen, all processes of burning would proceed with great rapidity and violence; but the nitrogen in the air amounts to nearly 80 per cent. of the whole atmosphere, and this inert gas moderates the intensity, and prolongs the duration, of the processes of combustion. Air is inhaled by all living animals, and changes which are essentially of the nature of burnings proceed in the tissues of animals; but along with the active element oxygen, animals also inhale the moderating element nitrogen, and the restraining action of the latter prevents the too rapid destruction of the tissues of the living organisms. At very high altitudes breathing becomes difficult; because the air at great heights is so attenuated that the work which must be done in inhaling a sufficient quantity of the gas which feeds the combustion-processes that are necessary accompaniments of the continuance of life is much greater than the work which need be done in order to obtain a sufficient supply of oxygen at ordinary levels. On the other hand, it has been found advantageous to supply patients suffering from certain diseases with pure oxygen, for inhalation, in place of ordinary air; because when a person is in a state of great exhaustion he is not able to exert sufficient muscular energy to enable him to gain oxygen from ordinary air in sufficient quantity to carry on those chemical changes that must be continued if life is to be prolonged.

These four elements, carbon, oxygen, hydrogen, and nitrogen, combine in most diverse proportions to form a vast multitude of compounds. The composition of every one of these compounds can be expressed by the chemical

formula $C_aO_bH_cN_d$. In this formula C means
an atom of carbon, O an atom of oxygen, H
an atom of hydrogen, and N an atom of nitrogen; and the small letters a, b, c, and d, represent whole numbers. As the atomic weights
of the four elements are these,—carbon $= 12$,
oxygen $= 16$, hydrogen $= 1$, and nitrogen $= 14$,
the formula also tells us that every compound
of these elements is composed of some whole
multiple of 12 parts by weight of carbon
united with some whole multiple of 16 parts
by weight of oxygen, some whole multiple of
one part by weight of hydrogen, and some whole
multiple of 14 parts by weight of nitrogen.
Some of the compounds we shall have to consider are composed of only two, or three, of the
four elements carbon, oxygen, hydrogen, and
nitrogen. But carbon is a constituent of every
one of them. Carbon is the characteristic element of the compounds which are produced by
the agency of living things, and of many compounds that are allied to these.

The fact that each one of an exceedingly great
number of compounds is a combination of carbon
with one, or more, of three other elements might
make us expect to find some reactions belonging
to these compounds, as a class, which should distinguish them, broadly, from all other compounds.
The study of the compounds of carbon justifies
this expectation. Looked at broadly, these compounds are characterised by an unwillingness to
enter into chemical reactions; they are slow
to change; many of their interactions drag,
pause before they attain completion, and stop at

resting-places on the road. For instance: all compounds of carbon with the elements hydrogen and oxygen can be burnt, by heating in oxygen, to produce water and carbonic acid gas; but if the processes of oxidation are conducted at, or about, the ordinary temperature of the air, a vast number of compounds can be produced many of which are of more complex composition than those which were the starting points of the reactions. If alcohol is warmed slightly in an open vessel, and a lighted taper is brought near the surface of the liquid, the vapour of alcohol takes fire, and the products of the burning are carbonic acid gas and water; but if oxygen is produced, somewhat slowly, in contact with alcohol, by heating the alcohol with some mixture of compounds which interact gradually to produce oxygen (for instance, with Condy's fluid and oil of vitriol), a substance called *aldehyde* is formed; and if this compound is subjected to a very gradual process of oxidation various compounds belonging to the class of sugars are produced. The composition of alcohol is expressed by the formula C_2H_6O; the formula C_2H_4O expresses the composition of aldehyde; and most of the sugars belong either to the class of compounds which have the composition $C_6H_{12}O_6$, or to the class which has the composition $C_{12}H_{22}O_{11}$. There is a comparatively simple compound of carbon and hydrogen called *acetylene*; the molecule of this compound is composed of two atoms of carbon united with two atoms of hydrogen—the formula of the compound is, therefore, C_2H_2. When this compound—it is a gas—is passed through a hot tube, among the products of the changes it undergoes is another, more complex, compound of carbon and hydrogen called *benzene* which has the composition C_6H_6. If benzene is burnt in the presence of plenty of air, or oxygen, water and carbonic acid gas, and these compounds only, are produced. But if benzene is gently warmed with *aqua fortis* (nitric acid), which is a compound that readily gives up part of its oxygen to other bodies, an oily liquid, smelling like oil of bitter almonds, is formed; this oily liquid is called *nitrobenzene*; its composition is $C_6H_5NO_2$. *Aniline*, $C_6H_5NH_2$, is made by warming nitrobenzene with iron and acetic acid; and by treating aniline with suitable oxidising reagents, the complex compound *rosaniline*, $C_{20}H_{21}N_3O$, is formed. Rosaniline is a colourless, crystalline solid; but it combines with acids to form a series of brilliantly coloured salts many of which are used for dyeing purposes.

These examples illustrate the statement that the interactions of the compounds of carbon, as a class, tend to the formation of compounds more complex than those wherewith the reactions begin. The reactions of compounds which do not contain carbon—when these bodies are contrasted with the compounds of carbon they are called *inorganic compounds*—generally proceed at once to their final goal; these reactions are direct, decisive, definite. The reactions of carbon compounds are hesitating, easy-going, dilatory; small changes in the conditions of a reaction are often accompanied by great changes in the compositions of the products; the reactions readily wander off into side-paths; the most trivial excuses for meandering

C

are accepted ; the mark of these chemical changes is, unwillingness to be done.

A closer examination of the reactions of the compounds of carbon enables us to say that the atoms of this element are ready to combine with one another, thereby producing groups of atoms —all atoms of carbon—which hold together firmly, and around which, as around nuclei or frameworks, may be built less stable combinations of other atoms. Such central nuclei of four, five, six, or more atoms of carbon may be compared to the square-set, massive, donjon of a castle ; and the subsidiary groups of other atoms which are attached to the central carbon nuclei may be likened to the hall, kitchens, buttery, and living rooms that are grouped around the central keep. If the castle was attacked, the outworks were carried first; the keep resisted to the last. So, when the molecule of a complicated carbon compound is attacked by chemical reagents, the outlying atomic groups are disintegrated or changed ; but in many cases the central nucleus of mutually linked carbon atoms remains intact.

Chemists picture to themselves different modes of linking of two or more atoms of carbon, in order that they may be able to think clearly and definitely about the connexions between the reactions of the carbon compounds and the compositions of their molecules. The existence of the molecules CH_4, CCl_4, CBr_4, CI_4, CCl_3Br, and the failures that have attended every attempt to isolate a molecule composed of more than four atoms of hydrogen, chlorine, bromine, or iodine united with a single atom of carbon, require us to say that the atom of carbon is *tetravalent*, or, to use another form of expression, that one atom of carbon can link to itself directly not more than four other atoms of any kind. Now it may reasonably be supposed that a pair of carbon atoms linked together may be able to hold directly to themselves either six, four, or two, atoms of hydrogen. The meaning of this supposition is better realised by presenting it in chemical symbols; using the conventional representation of atom-linking power by straight lines proceeding from the symbol of carbon. The structural formulæ of the three hydrocarbon molecules, each containing a pair of carbon atoms—the name *hydrocarbon* is given to all compounds of hydrogen and carbon only—are these :

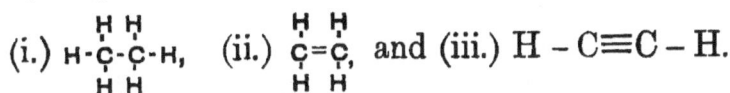

(i.) H-C-C-H,   (ii.) C=C, and (iii.) $H - C \equiv C - H.$

In the first case, one fourth of the atom-fixing power of each atom of carbon is thought of as used in holding together the two carbon atoms ; in the second case, one half, and in the third case, three-fourths, of the atom-fixing power of each carbon atom is supposed to be employed in binding the pair of carbon atoms together. The reactions of a molecule which contains a pair of carbon atoms linked in the manner that is represented by the symbol C – C will not be the same as the reactions of a molecule that contains a pair of carbon atoms linked in the manner represented by the symbol C = C; and the reactions of a molecule containing a pair of carbon atoms linked thus, C≡C, will differ from those of both of the other molecules; and these statements will hold good whatever be the atoms which are united with the pair of atoms of carbon. Inasmuch as the hydrocarbon C_2H_6 is called *ethane*, the hydrocarbon C_2H_4 is called *ethylene*, and the hydrocarbon C_2H_2 is called *acetylene*, it has become customary to speak of the *ethane-linking*, the *ethylene-linking*, and the *acetylene-linking*, of a pair of carbon atoms, in different molecules; one also speaks of a pair of *singly linked*, or a pair of *doubly linked*, or a pair of *trebly linked*, atoms of carbon.

This language is highly symbolical. The reader must guard himself against supposing that the symbols $C-C$ $C=C$ and $C\equiv C$ are intended to be presentments of the actual arrangements of the atoms of carbon in different molecules—one is obliged to think of all the parts of molecules as performing ordered movements—these symbols are only convenient and workable ways of stating facts in the language of a special hypothesis, which has itself grown out of the application of a general theory of the structure of matter to that class of natural occurrences called chemical reactions. One especially important fact is told by the symbols we are considering: six atoms of hydrogen or of other monovalent elements can enter into direct combination with the atomic group $C-C$; whereas only four atoms of hydrogen or of other monovalent elements can be directly combined with the group $C=C$; and only two atoms of hydrogen, etc., can be directly joined to the group $C\equiv C$.

Other modes of linking carbon atoms are conceivable ; and, in order to bring the facts of organic chemistry into one point of view, for the purpose of comparing and contrasting these facts, other modes of linking atoms of carbon must be presented in formulæ and used as working hypotheses. The most important of these modes of linking is that which is spoken of as the *benzene-linking*. The molecule of the hydrocarbon called. *benzene* is composed of six atoms of carbon united with six atoms of hydrogen ; the formula C_6H_6, therefore, expresses the composition of the molecule of this compound. A study of the reactions of benzene leads to the representation of the structure of the molecule of this compound, in terms of the hypothesis of

atom-linking, by the structural formula H C $\begin{smallmatrix} H & H \\ C-C \\ C-C \\ H & H \end{smallmatrix}$ C H .

The atom of carbon is here, as always, taken to be tetravalent, and the atom of hydrogen (as always) to be monovalent. In the *benzene-linking*, the six atoms of carbon are said to be alternately singly and doubly linked. We shall understand more fully what facts are sought to be conveyed by this expression when we have considered the reactions of benzene and its derivatives. Suffice it to say here that the reactions of benzene are altogether different from those of another hydrocarbon which has the same molecular composition as benzene.

As has been remarked already (p. 28), it is probable that the molecular weights of the three varieties of the element carbon, namely, diamond, graphite, and amorphous carbon, are not the same; that is to say, it is probable that the number of atoms of carbon bound together into a group which moves about as a whole is different in each case. But there is another possible way of representing the existence of three different carbon molecules in terms of the hypothesis of atom-linking: the three molecules may contain equal numbers of atoms of carbon, but these atoms may be arranged differently in each molecule; the molecule of one variety may be represented by such a symbol as $C - C - C - C$, the molecule of another variety by such a symbol as $\begin{smallmatrix} C\text{-}C \\ C\text{-}C \end{smallmatrix}$, and the molecule of the third form by such a symbol as $\begin{smallmatrix} C\text{-}C \\ C \\ C \end{smallmatrix}$. Or the differences between the three varieties of carbon may be thought of as connected with differences in the modes of linking of, say, a pair of carbon atoms: if this hypothesis is adopted, then one molecule may be represented by the symbol $C - C$, another by the symbol $C = C$, and the third by the symbol $C \equiv C$.

The atom of hydrogen is monovalent, the atom of oxygen is divalent, and the atom of nitrogen is trivalent; that is to say, an atom of hydrogen can bind to itself directly only a single other atom, an atom of oxygen can enter into direct union with two other atoms, and an atom of nitrogen can be linked directly with three other atoms. Now, consider the bearing of these statements on the question of the possibility of forming compounds by adding outlying groups of atoms to a pair of linked carbon atoms. Take a pair of singly linked atoms of carbon; $C - C$. It is possible, theoretically, to join to this group six atoms of hydrogen, or of other monovalent elements, but not more than six such atoms; the compounds C_2H_6, C_2H_5Cl, $C_2H_4Br_2$, etc., have been isolated. But suppose five atoms of hydrogen and one atom of oxygen to be united to the group $C - C$: as the oxygen atom is capable of uniting directly with two other atoms, the presence of this atom brings with it the possibility of adding another monovalent atom to the aggregate; the existence of such a molecule as C_2H_5OH becomes possible. This possibility is more clearly grasped if the structural formula is

employed :
$$\begin{array}{ccc} & H & H \\ H- & \overset{|}{C}-\overset{|}{C} & -O-H. \\ & H & H \end{array}$$
The compound C_2H_5OH

is *alcohol.* Now suppose that five atoms of hydrogen and one atom of nitrogen are combined with the pair of singly linked carbon atoms : as the atom of nitrogen can hold to itself directly three other atoms, the entrance of this nitrogen atom into the molecular building makes it possible, theoretically, to add two more monovalent atoms in order to complete the molecule. The structural formula of one such finished molecular

building is
$$\begin{array}{ccc} & H & H \\ H- & \overset{|}{C}-\overset{|}{C}-N & \overset{H}{\underset{H}{<}}. \\ & H & H \end{array}$$
The compound $C_2H_5NH_2$

is called *ethylamine.*

These illustrations throw some light on the statement, made on p. 34, that the atoms of car·

bon tend to combine together to form what may
be called the foundations of molecular aggregates,
and that the molecular structures are completed
by building on these foundations. When compounds whose molecules contain several atoms of
carbon are subjected to the action of moderately
active reagents, the tendency generally is towards
the occurrence of chemical changes in what may
be described as the outlying parts of the molecular
structures. For instance; there is a compound
called *dimethyl-benzene*, the composition and the
reactions of which lead us to give it the structural

formula . When this compound is

submitted to the action of oxidising reagents,
that is, compounds which react to produce
oxygen, the two outlying CH_3 groups (represented in the formula to the right and left of
the dotted lines) are changed into groups which
contain oxygen, and whose composition is ex-

pressed by the structural formula . The

compound produced is $C_6H_4(CO_2H)_2$; it is
called *phthalic acid*. When this phthalic acid
is heated with lime, benzene (C_6H_6) is obtained;
that is to say, the two outlying CO_2H groups
only are attacked, but the foundation of six
linked atoms of carbon remains unshaken. It
is only by the action of very energetic reagents,
aided by a high temperature, that the central
nucleus of six carbon atoms is broken down.
Another instance of the comparative ease with

which the side groups, or *side chains* as they are generally called, of molecular structures built on the foundation of six linked carbon atoms are attacked by reagents, while the central nucleus remains intact, is furnished by some reactions, the starting point of which is a compound called *diazobenzene nitrate*. The structural formula of

this compound is $H \cdot C \overset{\overset{\displaystyle H \ H}{C-C}}{\underset{\underset{\displaystyle H \ H}{C-C}}{}} C\text{-N-N-O-N} \overset{O}{\underset{O}{\langle}}$. When this

compound is boiled with water, nitrogen is given off, nitric acid (HNO_3) is formed and dissolves in the water, and a compound called *phenol* (more commonly known as *carbolic acid*), the composition of which is expressed by the formula $C_6H_5.OH$, is also produced; in other words, the whole of the side chain $N = N - NO_3$ is removed; another and simpler side chain, OH, is put in its place; but the central nucleus is untouched. If phenol is treated with iodine, the compound *tri-iodophenol*, $C_6H_2I_3.OH$, is formed; if phenol is caused to react with nitric acid, three compounds are formed, called *nitrophenol, dinitrophenol*, and *trinitrophenol*, and having the compositions $C_6H_4.NO_2.OH$, $C_6H_3.(NO_2)_2.OH$, and $C_6H_2.(NO_2)_3.OH$; if phenol is heated with concentrated sulphuric acid a compound called *phenolsulphonic acid* is formed, the composition of which is $C_6H_4.HSO_3.OH$; lastly, the products of distilling phenol with zinc dust are, zinc oxide and benzene C_6H_6. In none of these reactions is the central group of six linked carbon atoms broken down; the reactions consist in removing side chains and putting other groups in their places, or in putting groups of atoms, or single atoms, in the places of one or more of the atoms of hydrogen which are attached to the foundation of six linked atoms of carbon.

The examples of the reactions of carbon compounds we have been considering shew that the language wherein these reactions are recorded—and it is the only language that has been found capable of expressing the facts intelligibly—is steeped in the conceptions of the molecular and atomic theory. Unless a clear picture is kept in one's mind of the molecule as an aggregate of atoms arranged in an orderly manner, and held together by actions and reactions between its parts, the facts of organic chemistry become merely a dust heap of disconnected details. The conception of the fixed atom-linking powers of the different kinds of atoms, and that of the removal of a group of atoms from a molecule and its replacement by another group, or by a single atom, are perhaps the two most important molecular and atomic conceptions to be grasped clearly in trying to follow the element carbon in its wanderings from compound to compound.*

The compounds of carbon are slow to enter into deep-reaching chemical reactions ; they are ready to form more complex compounds ; and

* I would request the reader to refer to pp. 120 to 141, and 175 to 178, of *The Story of the Chemical Elements* for an elementary treatment of the notion of atom-linking, or atom-fixing, power ; and to pp. 169 to 175 of the same book for a brief consideration of the notion of the compound radicle, or atomic group.

also to break down to simpler compounds, but
only to a limited extent; they are characterised
by a kind of immobile plasticity; they are at
once stable and unstable; small alterations in
their conditions induce changes, but these
changes proceed leisurely with many stopping
places: while these statements are true, it is
also true that under the proper conditions the
compounds of carbon are decomposed with the
production of carbonic acid gas, and water, and
some simple nitrogenous compounds. The
compounds of carbon are evidently well suited
for the part they play in the life-processes of
plants and animals: their number is legion;
their properties vary much; every phase of the
life of an organism is marked by the production
or the disintegration of some of these compounds; changes in the conditions of the organism are accompanied by changes in the kind of
carbon compounds formed within it, and the
production of new compounds must tend to cause
a development, or a retrogression, of the organism in new directions. It was supposed at one
time that the carbon compounds which are
formed in living organisms could be formed only
by the agency of what was called "vital force";
but since thousands of such compounds have
been synthesised in the laboratory, that hypothesis, which was essentially unscientific, inasmuch as it did not help research into natural
occurrences, has been abandoned. The hypothesis of "vital force" does not aid us in
classifying and generalising the facts of organic
chemistry. It would be better not to employ

the expression *organic chemistry*, were it not that the more accurate phrase *the chemistry of the compounds of carbon* is too cumbersome. At the same time we ought not to forget that our knowledge of the compositions and reactions of the compounds of carbon has not enabled us to explain the phenomena of life.

CHAPTER III.

AN OUTLINE OF THE CLASSIFICATION OF SOME OF THE COMPOUNDS OF CARBON.

THE purposes of classification are these :—To place together those objects which are like, and to separate those which are unlike ; to discover and set forth general expressions of the similarities that are observed ; to aid the formation of clear conceptions of the characters of the objects classified ; and to help the memory to retain these conceptions.

The object of this book will be better served by attempting to classify a comparatively small number of compounds of carbon, than by allowing the attempt to range over the whole field. By far the greater number of the compounds of the element whose migrations we shall endeavour to follow are composed of carbon united with some, or all, of the three elements, hydrogen, oxygen, and nitrogen : although I propose to deal in the main only with compounds of these four elements, it will be necessary to make mention also of carbon compounds that contain chlorine, bromine, or iodine.

I shall first of all consider, briefly, the two oxides of carbon. The other compounds of carbon to which attention will be directed fall into two main classes ; those that more or less resemble paraffin, and those which on the whole are like benzene. The name *paraffinoid compounds* is given to the first class, and the name *benzenoid compounds* to the second class. The terms *fatty* and *aromatic* are much used, as synonymous with paraffinoid and benzenoid; because the former compounds include the common fats, and the latter include the aromatic bodies that are found in plants. The starting points from which the arrangements of the two classes of compounds proceed are, the hydrocarbon marsh gas for the first class, and the hydrocarbon benzene for the second class.

The systematic name of marsh gas is *methane*. From this compound, which is the simplest combination of carbon and hydrogen—its composition and molecular weight are expressed by the formula CH_4—are derived : first, *chloromethane* (CH_3Cl) and *chloroform* ($CHCl_3$), then *methylic alcohol* ($CH_3.OH$) (commonly known as wood spirit), then *methylic aldehyde* (CH_2O), *formic acid* (CH_2O_2), and *methylamine* ($CH_3.NH_2$). A similar series of compounds is derived from the simplest hydrocarbon formed by the union of hydrogen atoms with a pair of atoms of carbon : *ethane* (C_2H_6) forms *chloro-ethane* (C_2H_5Cl), *ethylic alcohol* (C_2H_5OH), *ethylic aldehyde* (C_2H_4O), *ethylic ether* ($[C_2H_5]_2O$), *acetic acid* ($C_2H_4O_2$), and *ethylamine* ($C_2H_5.NH_2$). Mention will be made of certain other compounds belonging to the classes of alcohols and acids ; and we shall also consider a few compounds derived from acids, especially certain *ethereal salts*.

The *sugars*, and the compounds allied to them called starches or *amyloses* will be glanced at.

From *benzene* (C_6H_6), which is the simplest member of the class of aromatic compounds, is derived *phenol* ($C_6H_5.OH$), *benzoic acid* ($C_6H_5.CO_2H$), and *salicylic acid* ($C_6H_4.OH.CO_2H$) ; also *nitrobenzene* ($C_6H_5.NO_2$), and *aniline* ($C_6H_5.NH_2$). The aromatic hydrocarbon *anthracene* ($C_{14}H_{10}$) is connected with the *alizarin colours*. Lastly, it will be advisable to consider, very slightly, the *alkaloids*, and *albumin*.

The division of the compounds of carbon into the two classes of fatty and aromatic compounds rests, primarily, on differences of properties. An examination of the reactions of members of each class shews that the differences in properties are accompanied by differences in composition, but that the latter differences are rather dissimilarities in molecular structure than in the relative quantities of the elements in the molecules of the two kinds of compounds. For instance ; there are two hydrocarbons whose composition is expressed by the formula C_6H_6. The compositions of these two compounds are identical, if the term *composition* is employed in the ordinary restricted sense. A molecule of one of the hydrocarbons combines very readily and rapidly with eight atoms of bromine to form a molecule of the compound $C_6H_6Br_8$: the other hydrocarbon molecule will not combine with more than six atoms of bromine, and the formation of the molecule $C_6H_6Br_6$ is accomplished only under the influence of direct sunshine and then very slowly. These and other reactions of the two hydrocarbons shew that one of them belongs to the class of fatty (or paraffinoid) compounds, and the other to the class of aromatic (or benzenoid) compounds. The only means we have of intelligibly representing the differences between the reactions of the two hydrocarbons is by supposing that the six atoms of carbon are arranged in essentially different ways in the two molecules, C_6H_6.

The consideration of the processes by which the changes are effected from the hydrocarbon methane (CH_4) to formic acid (CH_2O_2), or from the hydrocarbon ethane (C_2H_6) to acetic acid $(C_2H_4O_2)$, and the examination of the gradual modifications of properties and reactions which accompany these changes of composition, will furnish examples of what was said in the last chapter about the general character of the reactions of the compounds of carbon. The many stages that may be distinguished in the course of a chemical change between carbon compounds, and the unwillingness of these changes to rush to completion, will also be illustrated by following the string of reactions that begins with benzene (C_6H_6), and, for our purpose, ends with salicylic acid $(C_6H_4.OH.CO_2H)$. Such slight examinations as we shall be able to make of one or two sugars, of alizarin and the compounds allied to alizarin, and of the alkaloids, will serve to introduce the reader to the conception of molecular symmetry, and to give him a glimpse of the vast field of inquiry that is opening as the influences of the presence of atoms of carbon, and of their modes of linking with one another and with other atoms, on the properties and the functions of atomic aggregates, are being connected with the special phenomena which are exhibited by living organisms.

I hope also, to give illustrations of the employment in chemical industries of the reactions and properties of the compounds whose classification is being considered; and to show how large and extensive are many of the technical applications of chemical changes which would certainly appear to a superficial observer to be absolutely disconnected with anything that is "useful."

CHAPTER IV.

THE TWO OXIDES OF CARBON.

IF air is passed slowly through a heap of smouldering charcoal, coke, or coal, chemical changes take place, and that compound of carbon which is the chief product of these changes is composed of one atom of carbon united with one atom of oxygen. If a rapid current of air is driven over a small quantity of burning charcoal, coke, or coal, another oxide of carbon is produced, and this compound is formed by the union of one atom of carbon with two atoms of oxygen. The composition of the first compound is expressed by the formula CO, and that of the second compound by the formula CO_2. The former compound is called *carbon monoxide*, the latter, *carbon dioxide* or carbonic acid gas. Both compounds are colourless gases; both are poisonous: carbon monoxide poisons by removing oxygen from the blood, and forming a compound with a body called *hæmoglobin* which is the main constituent of the red corpuscles of the blood of vertebrate animals; carbon dioxide, called *choke-damp* by miners, poisons by cutting off the supply of oxygen and causing suffocation. Carbon monoxide takes fire when a light is brought near it, if air or oxygen is present, and burns with a pale blue-violet flame, producing carbon dioxide. The dioxide is not combustible, and the presence of a few parts of this gas in a hundred parts of air stops the burning of a candle. Carbon dioxide is about one and a half times heavier than air, bulk for bulk; if this gas is produced in a closed place where the air is at rest it accumulates in the lower portions of the enclosed space. Any specified volume of carbon monoxide weighs very slightly less than an equal volume of air. Carbon monoxide readily removes oxygen from many compounds of that element, forming carbon dioxide and some substance containing less oxygen than that from which the oxygen has been removed; for instance, most oxides of metals are deprived of their oxygen by heating them in a stream of carbon monoxide. Carbon dioxide, on the other hand, does not combine with oxygen; it has not been found possible to cause more than two

D

atoms of oxygen to combine with one atom of carbon. It is customary to express these facts, that carbon monoxide will combine directly with oxygen (and also with certain other elements) and that the dioxide will not combine directly with oxygen (or other elements), by saying that carbon monoxide is *an unsaturated compound*, and carbon dioxide is *a saturated compound*. Carbon monoxide is very slightly soluble in water; one pint of water at about 50°F. dissolves about ·027 of a pint of the gas. The solution is neither alkaline nor acid. Carbon dioxide is more soluble in water than the monoxide; one pint of water at 50°F. dissolves about 1·185 pints of the gas at the ordinary pressure of the air, and at 68°F. only about nine-tenths of a pint at the atmospheric pressure. The weight of the gas dissolved increases directly with the pressure. In other words; while one litre of water (about 1¾ pints) dissolves a quantity of carbon dioxide weighing 2·346 grams (about 36⅓ grains) at 10°C [50°F.], when the pressure on the surface of the water and the gas is the ordinary pressure of the atmosphere (equal to about 14¾ lbs. on the square inch), the same volume of water at the same temperature dissolves 2·346 × 2 = 4·692 grams (about 72⅔ grains) when the pressure is doubled, and 2·346 × 3 = 7·038 grams (about 109 grains) when the pressure is trebled. If the pressure on a solution of 4·692 grams of carbon dioxide in one litre of water is reduced to one half, 2·346 grams (that is, one half) of the carbon dioxide escape; and if the pressure on a solution of 7·038 grams of the gas in one litre

of water is reduced to one third, 4·692 grams
(that is, two thirds) of the gas escape. A solution of carbon dioxide in water has feebly acidic
properties.

Soda water is made by forcing carbon dioxide
into water by increasing the pressure on the surface of the gas and the water. It is customary to
bottle soda water at a pressure varying from 8 to
9 atmospheres ; that is to say, at a pressure from
8 to 9 times greater than the ordinary pressure
of the atmosphere.   As the ordinary atmospheric
pressure is equal to about $14\frac{3}{4}$ lbs. on the square
inch, soda water is bottled at a pressure equal to
from $14\frac{3}{4} \times 8 = 118$ lbs., to $14\frac{3}{4} \times 9 = 132\frac{3}{4}$ lbs., on
the square inch.   Now, one pint of water will
dissolve, at the ordinary temperature (say, 50°F.),
approximately 166 grains of carbon dioxide when
the pressure is equal to that of 8 atmospheres
($= 118$ lbs. on the square inch), and approximately 187 grains of carbon dioxide when the
pressure is equal to that of 9 atmospheres ($= 132\frac{3}{4}$
lbs. on the square inch); when the pressure is
reduced to the ordinary atmospheric pressure by
opening the bottle of soda water, about 130 grains
of carbon dioxide, equal to about $1\frac{2}{5}$ gallons of
the gas, will escape in the first case, and about
151 grains, equal to about 2 gallons of the gas,
will escape in.the second case.

Both carbon monoxide and dioxide have been
liquefied, and both have been frozen to white,
snow-like solids, by subjecting them to great
pressure at a very low temperature.   If carbon
dioxide gas is compressed it refuses to become
liquid unless the temperature is 30·9°C. [87·6°F.],

or lower than 30·9°C. At that temperature, a
pressure 73 times greater than the ordinary
pressure of the atmosphere (that is, 14¾ × 73 =
1076¾ lbs. on the square inch) is required to
change gaseous into liquid carbon dioxide. At
lower temperatures less pressure is needed to
effect the change. The temperature 30·9°C.
[87·6°F.] is called the *critical temperature* of
carbon dioxide. There is a critical temperature
for each gas, that is, a temperature above which
the gas cannot be caused to become liquid by
compressing it. The critical temperatures of different gases are very different; for instance, the
critical temperature of gaseous alcohol is about
240°C. [464°F.], and that of hydrogen is about
minus 234°C. [*minus* 389°F.]. This statement
means that gaseous alcohol can be liquefied by
pressure provided the temperature is lower than
464°F.; but that if it is wished to liquefy hydrogen
by compressing it, the gas must be cooled to at
least 389° below the zero of Fahrenheit's scale.

Liquid carbon dioxide is a marketable com-
modity; it is sold in strong steel bottles which
generally contain about 15 lbs. of the liquid
= about 12 cub. feet of the gas at the ordinary
temperature of the air. Liquid carbon dioxide
is used for extinguishing fires, for propelling
torpedoes, for obtaining cold by allowing the
liquid to evaporate, for aërating water, and for
other purposes. Small spheroidal vessels of steel
containing liquid carbon dioxide are now sold,
under the name of "sparklets," for making soda
water. One of the little vessels is placed in a
rubber washer at a short distance above the

water in a strong glass bottle; a sharp pin is fixed in the middle of the washer, and the " sparklet " is forced by a screw-cap on to this pin which perforates the material of the little vessel; the liquid carbon dioxide at once becomes gas, and this gas bubbles into the water wherein it dissolves.

Enormous quantities of carbon dioxide exist in the atmosphere. The total weight of this gas in the air is probably about 3 billion tons; more accurately, 10,000 volumes of country-air contain about 3 volumes of carbon dioxide. The quantity of this gas in the air varies slightly in different localities and at different seasons; there is slightly more in sea-air than in the air of country places, and country-air contains a little more carbon dioxide during the night than during the day. The air of towns contains more carbon dioxide than the air of country places : the average amount in 10,000 volumes of the air in the streets of London is 3·8 volumes, in the streets of Manchester, 4 volumes, and in the streets of Glasgow, 5 volumes, while the average amount in 10,000 volumes of country-air is 2·99 volumes. Air taken from the tunnels of the Metropolitan railway was found to contain from 14 to 15 volumes of carbon dioxide in 10,000 volumes of the air. As fogs interfere with the circulation of the air and with the mixing of the various constituents by diffusion, it is to be expected that the proportionate quantity of a gas which is being poured every moment into the atmosphere of every town, as a product of the burnings of fuel and the breathings of men and animals, will be

very much altered by the occurrence of fogs. This
expectation is justified by measurements of the
quantity of carbon dioxide in town-air during
fogs ; the quantity of that gas sometimes amounts
to 10 volumes per 10,000 of foggy air. Inasmuch
as 100 volumes of the expired breath of human
beings contain about $4\frac{1}{2}$ volumes of carbon dioxide, it is evident that the atmosphere of rooms
wherein many people are congregated must soon
become rich in this poisonous gas, unless some
efficient method of ventilation is adopted to remove the carbon dioxide and replace it by fresh
air. A school-room, or lecture-room, is generally
supposed to be efficiently ventilated if the amount
of carbon dioxide does not exceed 6 or 8 volumes
per 10,000 volumes of air ; in very many school-
rooms the quantity of carbon dioxide amounts to
10 volumes per 10,000 ; it frequently rises to 15
volumes, and in not a few cases to 20, or 25, or even
to 35 volumes per 10,000 volumes of air ; in some
schools in Austria as much as 55 volumes of this
gas have been found in 10,000 volumes of the air
of the rooms. In some cases, for instance in
schools at Aberdeen, Dundee, and Edinburgh, it
has been shown that the highest Government
grants, per scholar, are earned by those children
who attend the best ventilated schools ; in the
schools at Sheffield, on the other hand, no connexion could be traced between the ventilation of
the schools and the amount of the grants earned.

Carbon dioxide is produced in vast quantities
by the burning of coal, coke, charcoal, oil, candles,
and all fuel and illuminants which contain carbon.
This gas is found in almost all natural waters ;

it is one of the products of the putrefaction and decay of animal and vegetable matter; and it is produced during the fermentation of sugar-containing materials. When chalk is heated, it is decomposed into lime and carbon dioxide. This process is conducted, on a large scale, in lime-kilns. A quantity of coal is thrown into the kiln, then a cartload of chalk broken into pieces, then more coal, then more chalk, and so on; the coal is ignited, and the heat produced causes the separation of the chalk into carbon dioxide which escapes at the top of the kiln, and lime which sinks to the bottom. The process is generally made continuous, by throwing more chalk and fuel into the kiln at the top, while the lime is removed at the bottom from time to time. The reverse process to that which occurs in a lime-kiln is accomplished by passing carbon dioxide over lime; the two compounds unite, and chalk is formed.

Pure carbon dioxide is usually prepared by decomposing pure calcium carbonate (chalk and marble are more or less pure calcium carbonate) by diluted hydrochloric acid, washing the gas by passing it through water, drying it by causing it to bubble through concentrated sulphuric acid, and collecting it in dry vessels. Pure carbon monoxide may be prepared by burning a large quantity of pure carbon in a slow stream of oxygen, and passing the gaseous products of the reaction, in a slow current, through a solution of potash which absorbs the carbon dioxide but not the monoxide; the gas that is not absorbed by the potash solution is washed, dried, and collected. The monoxide is more generally prepared by decomposing oxalic acid by hot concentrated sulphuric acid, passing the gases produced, which consist of the two oxides of carbon, through several bottles containing a solution of potash, and then washing and drying that gas which is not absorbed by the potash.

Carbon dioxide can be decomposed into its elements by passing the gas over burning magnesium; the oxygen of the dioxide combines with the magnesium, forming magnesia, and carbon appears as a black solid. A process somewhat like this is brought about by the green parts of living plants aided by the sunlight. Plants absorb carbon dioxide, from the air and the rain, and separate it into its constituent elements; much of the oxygen which is thus produced is breathed out by the plants into the atmosphere, while almost the whole of the carbon is retained and used by the plants in building up their tissues. It is not to be supposed that the living plant actually separates the carbon dioxide into carbon and oxygen, *and then* brings about the combination of the separated carbon with other elements; rather, the compounds in the green parts of the plant, aided by sunshine and moisture, react with the carbon dioxide which the plant has absorbed, and the products of these reactions are, various more or less complex carbon compounds which remain in the plant, and oxygen which is returned to the atmosphere. It is also to be remembered that only a portion, although a large portion, of the oxygen of the absorbed carbon dioxide is sent

back into the atmosphere; some of it is used
by the plant, just as the carbon is used, in
building up its tissues. These processes occur
under the influence of sunshine; the reactions
which take place in the living plant during the
night are broadly similar to those that occur in
a living animal, and consist, in the main, in
changes wherein carbon dioxide is produced and
exhaled by the plant.

When a stream of carbon dioxide is passed
through a heap of red hot charcoal, or coke,
coal, or other carbonaceous fuel, half of the
oxygen of the dioxide is taken away by the
hot carbon, and carbon monoxide is produced.
The changes of composition that occur may
be expressed in a *chemical equation* thus:
$C + CO_2 = 2CO$. The gas which passes away
from the upper surface of the heap of burning
fuel consists chiefly of carbon monoxide, mixed,
perhaps, with a little of the dioxide which has
escaped the deoxidising action of the carbon of
the fuel; if the gas issues from the surface of
the burning fuel into the air, the carbon monoxide in the hot gas combines with oxygen in
the air and is burnt to carbon dioxide. The
change that takes place at the surface of the
burning heap of fuel may be expressed in an
equation thus: $CO + O = CO_2$. At the beginning of this chapter it was stated that carbon
dioxide is produced by passing plenty of air,
or oxygen, into burning charcoal, coke, or coal.
Remembering these reactions, let us consider the
chemical changes that take place in a coal fire
when the coal has become red hot. Air is con-

stantly entering at the bottom and sides of the grate : some of the oxygen in the air seizes the carbon in the outer and lower layers of coal and combines with it to produce carbon dioxide ; as this carbon dioxide passes through the glowing coal in the inner part of the grate it is deprived of half of its oxygen by the red hot carbon, and carbon monoxide is formed ; finally, as the carbon monoxide leaves the surface of the burning fuel it mixes with plenty of air, and as both the air and the monoxide are very hot, the carbon monoxide is burnt to carbon dioxide which passes by the chimney into the air outside the house. When a large fire is burning in a grate, and the coal has become red hot throughout, and has ceased to give off smoke and little puffs of gas which take fire and burn with some brilliancy, a pale bluish flame may be seen playing fitfully over the upper surface of the mass of glowing fuel ; this flame is the accompaniment of the burning of carbon monoxide to dioxide that is taking place where the monoxide, produced in the centre of the fuel, meets the oxygen in the air of the chimney.

I wish to direct attention to one other reaction whereby a mixture of the two oxides of carbon is produced. When steam is passed over such carbonaceous materials as coke, anthracite coal, or charcoal, the steam is robbed of its oxygen, and hydrogen and oxides of carbon are formed. By regulating the temperature, it is possible to produce either carbon dioxide or carbon monoxide, or a mixture of these gases. The principal reaction that occurs between steam and hot carbon at temperatures lower than 500°C. [932° F.] is expressed by the following chemical equation : $2H_2O + C = CO_2 + 4H$; at a temperature towards 1000°C [1832°F.] the same quantity of steam reacts with twice as much carbon as at the lower temperatures, and the products are hydrogen and carbon monoxide ($2H_2O + 2C = 2CO + 4H$.) As both carbon monoxide and hydrogen are gases that are easily burnt, and as the burning of these gases is accompanied by the production of much heat but very little light, an inflammable gas of great heating power, but of no use as an illuminant, may be produced by passing steam over extremely hot carbon. The product of this reaction is known as *water-gas*. The changes that occur when water-gas is burnt in air are these : (i) $4H + 2O = 2H_2O$, (ii) $2CO + 2O = 2CO_2$. The heating power of water-gas is rather less than half that of an equal volume of coal-gas. On the other hand, the volume of air required for the complete combustion of water-gas is less than one third of that required for the perfect burning of an equal volume of coal-gas, and the flame of water-gas is absolutely free from smoke ; for these reasons, water-gas is very useful for obtaining high temperatures when a 'clear' heat is required, as, for instance, in welding metals, burning porcelain, and melting glass. Some years ago it was noticed that water-gas which had been kept compressed in steel cylinders for some time emitted much light when it was burnt, and that iron was deposited from the burning gas. Investigation proved that part of the carbon monoxide in the water-gas had combined with some of the iron of the cylinders, producing a volatile compound of iron and carbon monoxide, and that this compound was decomposed when the gas was burnt, with the deposition of particles of oxide of iron which became red hot and evolved much light.

The discovery that carbon monoxide combines with iron to form a volatile substance was preceded by the discovery of a volatile compound of carbon monoxide and nickel. When finely powdered oxide of nickel is heated to bright redness in a stream of hydrogen, the oxygen is removed and very finely divided nickel remains; if this nickel is allowed to cool gradually in contact with carbon monoxide a gaseous compound of the nickel with the monoxide is produced; and when this gaseous compound is heated strongly it is decomposed into nickel, which is deposited as a lustrous solid, and carbon monoxide which passes away as a gas. These reactions suggested a new method for extracting nickel from its ores. Most nickel ores contain the metal combined with arsenic, or sulphur; when these ores are roasted in air nickel oxide is produced; if the oxide is then deoxidised by heating in a stream of hydrogen, the metal thus formed is allowed to cool in a slow current of carbon monoxide, and the gaseous compound of the nickel and the carbon monoxide is heated, pure nickel is obtained. This process has been used on the large scale recently; it is simpler and more expeditious than the older methods for extracting nickel from its ores; and, considering the large quantity of nickel now used for nickel plating, the new process is likely to come into favour.

Mention has been made of the deoxidising action of carbon monoxide. When this gas is passed over red hot oxide of copper, the oxygen of the oxide is removed, and carbon dioxide and copper are produced. Many other oxides of metals are reduced, or deoxidised, by heating them in contact with carbon monoxide, and the metals are obtained. The deoxidising action of carbon monoxide is made use of, on a very large scale, in the manufacture of iron from ironstone in blast furnaces. Charges of iron oxide and coke are thrown into the furnace, which resembles an enormous lime-kiln, the fuel is ignited, and hot air is blown in near the bottom of the furnace ; the carbon of the coke is burnt by the oxygen of the air, chiefly to carbon dioxide; the carbon dioxide is robbed of half of its oxygen by the hot carbon, with which it comes into contact at a little distance above the point where the air-blast is sent in, and carbon monoxide is produced ; the carbon monoxide reacts with the oxide of iron to produce carbon dioxide and iron ; the iron sinks, and melts, and is run from the furnace into moulds ; the ascending carbon dioxide is reduced to monoxide by contact with fresh layers of hot carbon ; more iron oxide is deprived of its oxygen, and more carbon dioxide is formed ; the dioxide is again reduced to monoxide which reacts once more with oxide of iron ; finally, carbon monoxide, mixed with some dioxide, passes off at the mouth of the furnace. Many other reactions occur in the blast furnace; for our purpose it is sufficient to notice one of these. The hot carbon of the coke and the oxide of iron interact to produce iron, and carbon monoxide; so that, besides what is formed by the combustion of the fuel, a supply of the deoxidising agent is produced by the direct action of the fuel on the oxide of iron.

In this chapter we have followed a few of those wanderings of carbon wherein the element does not venture very far afield. The partner of carbon in the journeyings which we have noticed has been the element oxygen. We have seen an atom of carbon joining company with a pair of atoms of oxygen, and the triplet swinging off into the atmosphere. We have seen the meeting of the molecule of carbon dioxide with a molecule of lime, and the production, by their union, of a molecule of chalk. We have followed the splitting of this molecule of chalk, in a lime kiln, into the two simpler molecules which coalesced to form it, and the return to the atmosphere of the carbonic acid gas which had been for a time imprisoned in the soil by reason of its temporary union with lime. We have pictured to ourselves the seizure of the molecule of carbon dioxide by a plant, the dissolution, by the combined action of the green parts of the plant and the sunshine, of the partnership of the atoms of carbon and oxygen, and the incorporation of the atom of carbon whose wanderings we have been following into the substance of the plant. It is not difficult to imagine the consumption of the plant by an animal, the union in the body of the animal of

the carbon atom with two atoms of oxygen, and
the escape into the atmosphere of the molecule of
carbonic acid gas thus formed : nor would it be
less easy to follow this molecule as it is carried,
by the inrushing stream of air, into a brightly
burning fire of coals, to see one of the atoms of
oxygen torn from the other atoms of the molecule
by a red hot atom of carbon, and to behold the
restoration to the molecule of carbon monoxide
of another atom of oxygen, in place of that which
the carbon removed, at the moment when the
molecule leaves the glowing fire and comes into
contact with the oxygen in the cooler air.
Instead of following the molecule of carbon
dioxide into a coal-fire, we might picture its
entry into a blast furnace ; and, after half of its
oxygen had been removed by one of the crowd of
rushing carbon atoms which it encountered
there, we might be spectators of the seizure by
the half satisfied molecule of carbon monoxide
of an atom of oxygen which had long been
quietly united with iron, and the formation, once
more, of a molecule of carbonic acid gas. And,
finally, we may follow the molecule of carbonic
acid gas as it is again captured, and torn asunder,
by a growing plant ; and then, passing over tens
of thousands of years, we may be mental spectators of the digging from the earth of a piece
of coal that has been produced by the very
gradual decay, in the absence of air and light,
of the plant which long ago had built into its
structure the atom of carbon whose migrations
we have been witnessing. In this piece of coal
we recover our atom of carbon unworn and un-

changed, in every respect the same as when it
started on its wanderings it may be a million
years ago.

CHAPTER V.

MARSH GAS, AND CERTAIN COMPOUNDS
DERIVED THEREFROM.

IF the mud at the bottom of a pool which contains the remains of dead plants is stirred,
bubbles of gas rise to the surface. If a wide-
mouthed bottle is filled with water, and inverted
in the pool so that the rising bubbles of gas pass
into the bottle, the water is driven out of the
bottle, and its place is taken by the gas. Suppose
that a bottle has been filled with this gas, and
that a lighted taper is brought near the mouth
of the bottle ; the gas takes fire, and burns with
a tolerably luminous flame. Now let another
bottleful of this *marsh gas* be mixed with its own
volume of the gaseous element chlorine (which is
a yellow, very suffocating, gas obtained from
hydrochloric acid, or from bleaching powder),
and let the mixture be placed in sunlight ; a reaction proceeds slowly, and after some hours both
the marsh gas and the chlorine have disappeared,
and in their place have been produced hydrochloric acid gas, and a pleasantly smelling gaseous
compound, called *chloromethane*, which burns with
a green-edged flame when a lighted taper is
brought near it. Now let some of this chloromethane be dissolved in an aqueous solution of

caustic potash, let the liquid be boiled for several days (in an apparatus so arranged that any volatile products of the reaction are condensed and caused to flow back into the liquid), and let the contents of the vessel then be distilled ; a colourless, mobile liquid with a spirituous odour, boiling at about 64° C. [147° F.], is obtained. This liquid is called *methylic alcohol.* If this alcohol is oxidised very slowly and carefully, a portion of it is converted into *formic aldehyde* ; and if the oxidation is carried farther, a transparent, colourless liquid, with a sour taste, called *formic acid,* is produced. If methylic alcohol is heated with salammoniac, a colourless gas is formed which smells like ammonia. This gas is called *methylamine.*

We have now to examine these excursions of carbon, from the compound called marsh gas to that named methylamine. The compositions of the compounds of carbon with which we have to deal are expressed by the following formulæ :— marsh gas, CH_4 ; chloromethane, CH_3Cl ; methylic alcohol, CH_4O ; formic aldehyde, CH_2O ; formic acid, CH_2O_2 ; and methylamine, CH_5N. We shall find our main interest in this series of compounds in an attempt to trace the relations between the reactions of these bodies, and then to express, or at least to suggest, these relations of function by formulæ which exhibit relations of molecular structure. Marsh gas— the systematic name of this compound is *methane* —is chlorinated in sunshine ; one of the products of this reaction (for several compounds are produced) is kept in contact with a boiling, aqueous

E

solution of potash for some days; a portion of the methylic alcohol that is slowly formed is caused to react, very gradually, with oxygen; the formic aldehyde which is the product of this change is oxidised to acetic acid; finally, the rest of the methylic alcohol is boiled with salammoniac, and methylamine is obtained. The following chemical equations present statements of the changes of composition which occur in these five reactions :—

(i) $CH_4 + 2 Cl = CH_3Cl + HCl.$
(ii) $CH_3Cl + KOH = CH_4O + KCl.$
(iii) $CH_4O + O = CH_2O + H_2O.$
(iv) $CH_2O + O = CH_2O_2.$
(v) $CH_4O + NH_4Cl = CH_5N + H_2O + HCl.$

What are the characteristic chemical properties of the six compounds beginning with CH_4 and finishing with CH_5N? The chief reaction of chloromethane (CH_3Cl) which interests us at present is that with potash solution which has been mentioned already, and is represented in equation (ii) above. The mere presentation of the compositions of methane and methylic alcohol by the formulæ CH_4 and CH_4O seems to show that the statement made in former chapters that an atom of carbon never directly unites with more than four other atoms, is not true. But, is the atom of carbon in the molecule CH_4O in *direct* union with all the other atoms of the molecule? The form of this question must be changed before an answer can be hoped for by experimental inquiry. The question must be put in some such words as these :—are the reactions of methylic alcohol in keeping with the hypothesis that the atom of carbon is directly linked to all the other atoms in a molecule of this compound ?

The metal potassium (caustic potash is a compound of this metal with oxygen and hydrogen) dissolves in methylic alcohol, forming the compound CH_3KO; however great a quantity of potassium is added, no other compound is produced. This reaction is expressed in molecular and atomic language by saying that an atom of the metal potassium turns out one atom, and only one atom, of hydrogen from the molecule of methylic alcohol; hence, there is some difference of function between one of the hydrogen atoms and the other three hydrogen atoms in the molecule CH_4O. When hydrochloric acid gas (HCl) is passed into methylic alcohol, in the presence of some substance which very readily combines with water but does not interact with methylic alcohol, chloromethane and water are formed; the following equation expresses the change :—

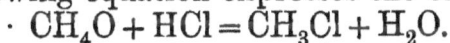

$$CH_4O + HCl = CH_3Cl + H_2O.$$

In this reaction, the atom of oxygen in a molecule of methylic alcohol has been removed along with one atom of hydrogen, and one atom of chlorine has been put in the place of the oxygen and hydrogen atoms taken away. Here again a difference is shewn between the part played, in the molecule, by one of the atoms of hydrogen and the parts played by the other three hydrogen atoms. In both of the reactions described, three atoms of hydrogen remain in combination with the atom of carbon in the molecule CH_4O; in one reaction an atom of hydrogen is taken out of the molecule (in exchange for an atom of potassium), and in the other reaction an atom of hydrogen and also the atom of oxygen are removed (in exchange for an atom of chlorine). We do not know what are the relations to one another of the atoms in the molecule CH_4O; but, using one of the rough, crude, symbols with which we are compelled to work at present, we can represent the inter-atomic relations in this molecule by a structural formula, which certainly advances our knowledge, because it is an instrument that helps us to conduct further inquiries. The structural formula suggested for the molecule of methylic alcohol by the reactions that have

been described is
$$
\begin{array}{c}
\text{H} \\
| \\
\text{H-C-O-H.} \\
| \\
\text{H}
\end{array}
$$
This formula repre-

sents one atom of hydrogen in direct union with the only atom of oxygen in the molecule; it represents the other three atoms of hydrogen in direct union with the carbon atom of the molecule, and as all related in the same way to the rest of the molecule. The structural formulæ of the products of these reactions that have been described would be these :—

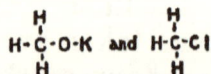

$$
\begin{array}{cc}
\text{H} & \text{H} \\
| & | \\
\text{H-C-O-K} \quad \text{and} \quad & \text{H-C-Cl} \\
| & | \\
\text{H} & \text{H}
\end{array}
$$

What we know of the reactions of other compounds of potassium, and of other compounds of chlorine, is in keeping with the assertion implied in these two formulæ that an atom of potassium, and an atom of chlorine, directly links to itself a single other atom.

Let us make another hypothesis. Let it be assumed that the four atoms of hydrogen and also the atom of oxygen are directly linked to the atom of carbon in the molecule of methylic alcohol: the structural formula of this molecule will then be $H-\overset{\overset{H}{|}}{\underset{\underset{H}{|}}{C}}-O$.

But if this is a correct translation into a language we can use as an instrument of thought, of the relations of the atoms that form the molecule CH_4O, why is there a difference between the function of one of the hydrogen atoms and the functions of the other three, and why cannot the atom of oxygen be removed, and replaced by chlorine, without carrying with it one of the atoms of hydrogen? Moreover, the symbol which represents the atom of carbon in direct union with all the other atoms of the molecule CH_4O, runs counter to what we know regarding the atom-fixing power of this element. If it is possible for a single atom of carbon to link directly to itself five other atoms, then we must rebuild the whole edifice of organic chemistry; wondering the while at the sight of a stable and stately building resting on nothing.

We must come back, then, to the symbol $H_3C.OH$, which is a less cumbrous form of the symbol $H-\overset{\overset{H}{|}}{\underset{\underset{H}{|}}{C}}-O-H$, as a working representation of the structure of the molecule of methylic alcohol. Is this structural formula in keeping with the changes that occur when methylic alcohol is oxidised? To answer this question, let us look at the processes whereby methylic alcohol is changed, first to formic aldehyde (CH_2O), and then to formic acid (CH_2O_2), and at the reactions of these two compounds. The first step of the change may be accomplished by passing a current of air charged with vapour of methylic alcohol over finely divided platinum, and then cooling the gases; in this way a solution of formic aldehyde in methylic alcohol is obtained. Finely divided platinum absorbs large volumes of many gases; in the present case it absorbs oxygen from the air, and also the vapour of methylic alcohol, and brings these two into such close contact that a chemical change occurs. Formic aldehyde removes oxygen from many compounds of that element, and is thus converted into formic acid, which is a colourless liquid with the sour taste and corrosive action on the skin that characterise the acids as a class. Formic acid, CH_2O_2, reacts with metals to form salts. All these salts— called *formates*—contain one atom of carbon, one atom of hydrogen, and two atoms of oxygen, or some whole multiple of one atom of carbon, one of hydrogen, and two of oxygen, in their molecules; in other words, they all consist of metal united with the group of atoms CHO_2, or with $2CHO_2$, or, generally, with $nCHO_2$, where n is a whole number (not greater than 4). The composition of the formates may be expressed by saying that only one atom of hydrogen in the molecule of formic acid (CH_2O_2) can be replaced by an atom of a metal. But this is equivalent to the statement that one of the two atoms of hydrogen in the molecule CH_2O_2 performs a function different from that of the other atoms ; one of the atoms of hydrogen seems to be more firmly held to the carbon atom than the other atom of hydrogen is held. We found the struc-

tural formula $\overset{\text{H}}{\underset{\text{H}}{\text{H-C-O-H}}}$ to be a fair expression of

the reactions of methylic alcohol. The structural

formula $\underset{\text{O}}{\overset{\text{H-C-O-H}}{\parallel}}$ expresses the reactions of

formic acid ; and the formula $\underset{\text{O}}{\overset{\text{H-C-O-M}}{\parallel}}$ expresses

the reactions of formates, where M is an atom of such a metal as potassium, sodium, or silver. In these formulæ one, and only one, atom of hydrogen is represented in direct union with the atom of carbon. The formulæ are in keeping with reactions by which the acid and its salts are prepared ; for instance, potassium formate is produced by passing carbon monoxide (CO) over moist caustic potash (KOH), and the acid is formed by the action of an electric discharge on a mixture of carbon dioxide and hydrogen. Both of these reactions are in keeping with the formulæ given above : thus,

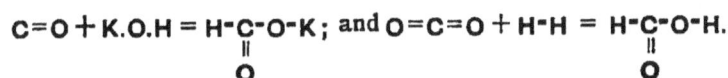

$$\text{C=O + K.O.H} = \underset{\text{O}}{\overset{\text{H-C-O-K}}{\parallel}} ; \text{and} \quad \text{O=C=O + H-H} = \underset{\text{O}}{\overset{\text{H-C-O-H}}{\parallel}}.$$

Moreover, formic acid readily decomposes to carbon dioxide and hydrogen ; and it is very easily oxidised to carbon dioxide and water. These two reactions find simple expressions by

using the structural formula given to the acid:
thus,

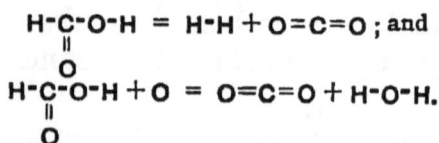

$$\text{H-C-O-H} = \text{H-H} + \text{O=C=O} ; \text{and}$$
$$\overset{\underset{\text{O}}{\|}}{}$$

$$\text{H-C-O-H} + \text{O} = \text{O=C=O} + \text{H-O-H.}$$
$$\overset{\underset{\text{O}}{\|}}{}$$

If, then, the reactions of formic acid are suggested by picturing the arrangement of the
atoms in a molecule of this compound by the

formula $\begin{matrix} \text{H-C-O-H} \\ \| \\ \text{O} \end{matrix}$; what structural formula will

best present the reactions of formic aldehyde in
terms of the arrangement of the atoms of the
molecule CH_2O ? The molecule of formic aldehyde contains an atom of oxygen less than the
molecule of formic acid; is it the atom of oxygen
in direct union with the carbon atom only, or
the atom of oxygen in direct union with atoms of
carbon and hydrogen, which is added to the molecule when the aldehyde is oxidised to the acid ?
In other words; which of the two formulæ

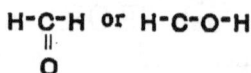

$$\begin{matrix} \text{H-C-H} \\ \| \\ \text{O} \end{matrix} \text{ or } \text{H-C-O-H}$$

better expresses the reactions of formic aldehyde ?
We found that methylic alcohol reacts as if the
atoms were arranged in a manner which can be
roughly represented by the structural formula
$H_3C - O - H$; we also found that when this
compound is heated with hydrochloric acid, the
oxygen atom and an atom of hydrogen are removed, their place is taken by an atom of
chlorine, and the compound $H_3C.Cl$ is produced.
Now, if the arrangement of the atoms in the

molecule of formic aldehyde is that pictured by the formula H - C - O - H, one might reasonably expect that hydrochloric acid would react with this compound (if it reacts at all) to form H - C - Cl. From the reactions of other compounds we learn that, when oxygen only is removed from a carbon compound and its place is taken by chlorine, two atoms of chlorine always take the place of a single atom of oxygen; hence, if formic aldehyde has the formula $\overset{\text{H-C-H}}{\underset{O}{\|}}$, we might expect hydrochloric acid to react (if it reacts at all) to produce the compound $\overset{\text{H-C-H}}{\underset{Cl_2}{\|}}$.

Neither oxygen alone, nor oxygen and hydrogen, can be removed from formic aldehyde by the action of hydrochloric acid; but if pentachloride of phosphorus (PCl_5) is employed—and this is a compound which reacts with many compounds of carbon to put chlorine in the place of oxygen, or of oxygen and hydrogen—oxygen only is removed, and two atoms of chlorine are put in the place of the atom of oxygen which is taken away from the molecule of formic aldehyde. In other words, we do not obtain H - C - Cl, but $\overset{\text{H-C-H}}{\underset{Cl_2}{\cdot}}$ Hence, we conclude that the structural formula to be assigned to formic aldehyde is $\overset{\text{H-C-H}}{\underset{O}{\cdot}}$.

The relations which an examination of their reactions shews to exist between the three compounds methylic alcohol, formic aldehyde, and formic acid, may, then, be conceived as relations between the arrangements of the atoms which form the molecules of these compounds by using the following formulæ :—

$$\text{H-C-O-H} \qquad \text{H-C-H} \qquad \text{and} \qquad \text{H-C-O-H}$$

Methylic alcohol.　Formic aldehyde.　　　　　Formic acid.

There remains the compound methylamine, CH_5N. What are the relations of this compound to methane, methylic alcohol, formic aldehyde, and formic acid? Methylamine is a colourless gas, which becomes a liquid at a temperature considerably under the freezing point of water; it smells like ammonia; like ammonia, it is very soluble in water, forming a solution which is very caustic (as a solution of ammonia is); this solution precipitates the hydroxides of many metals from solutions of salts of these metals, a reaction that is brought about also by a solution of ammonia. Methylamine combines with acids and forms salt-like compounds; and in each case the salt-like compound is composed of the methylamine and the acid, and not only of certain constituent portions of the methylamine and the acid. Ammonia also combines with acids, producing salts that are composed of one, two, or more molecules of ammonia united to one, two, or more molecules of the reacting acid. The reactions of methylamine are, evidently, like the reactions of ammonia. Now, similarity of reactions accompanies similarity of composition; using the word composition to include molecular structure. Hence the compositions of methylamine and ammonia are, probably, similar. When the formulæ of the two compounds are placed side by side, NH_3 and CH_5N, no marked likeness is forced on our notice. Let us turn back to the structural formulæ which were found good working representations of the reactions of methane (or marsh gas), chloromethane, and methylic alcohol; these formulæ are

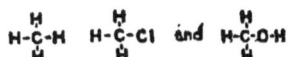

$$\begin{array}{ccc} \overset{H}{\underset{H}{H-C-H}} & \overset{H}{\underset{H}{H-C-Cl}} & \text{and} \quad \overset{H}{\underset{H}{H-C-O-H}} \end{array}$$

The second compound is derived from the first by replacing an atom of hydrogen by an atom of chlorine, and the third from the first by replacing an atom of hydrogen by an atom of oxygen and an atom of hydrogen directly linked together; the molecules of chloromethane and methylic alcohol are represented as each containing an atom of carbon directly united with three atoms of hydrogen. Now, suppose that the reaction between methylic alcohol and salammoniac, whereby methylamine is formed, is stated, in structural formulæ, in this way

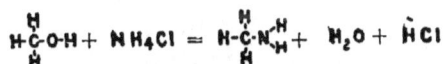

$$\overset{H}{\underset{H}{H-C-O-H}} + NH_4Cl = \overset{H}{\underset{H}{H-C-N}}\overset{H}{\underset{H}{}} + H_2O + HCl$$

then, the molecule of methylamine, also, is represented as containing an atom of carbon directly joined to three atoms of hydrogen; and, farther, the molecule of methylamine is represented as derived from a molecule of ammonia by replacing an atom of hydrogen by the group of atoms CH_3. The relation which is here asserted to exist between the molecular structures of am-

monia and methylamine becomes more evident
if the two formulæ are compared :—

$$\begin{matrix} H \\ \\ H \end{matrix} \hspace{-0.3em} N\text{-}H \qquad\qquad \begin{matrix} H & H \\ & \\ H & H \end{matrix} \hspace{-0.3em} N\text{-}C\text{-}H$$

Ammonia.   Methylamine.

But is the structural formula given to methylamine in keeping with other reactions of this
compound ? When hydriodic acid (HI) reacts
with ammonia, a compound of the two (called
ammonium iodide) is produced ; the reaction
may be thus expressed in an equation,

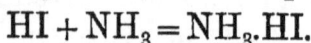

$$HI + NH_3 = NH_3.HI.$$

There is a compound called iodomethane, similar
to chloromethane, the composition of which is
given by the formula $(H_3C)I$; it is a compound
of an atom of iodine with the group of atoms
CH_3 (the bracket is used to shew this clearly).
If this compound reacted with ammonia what
should we expect to be formed ? I think we
might reasonably expect this reaction to occur,

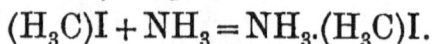

$$(H_3C)I + NH_3 = NH_3.(H_3C)I.$$

That is, we might reasonably expect a compound to be produced similar to that which is
produced by the combination of HI and NH_3.
In place of using HI we are using $(CH_3)I$; and,
therefore, in place of getting $NH_3.HI$ we expect
to get $NH_3.(CH_3)I$. Iodomethane and ammonia
combine to form NCH_6I. But is NCH_6I properly represented by the structural formula
$NH_3.(CH_3)I$? Turn back to the compound of
ammonia and hydriodic acid, NH_3HI : when
this compound is heated with potash, there are

produced ammonia, iodide of potassium, and water ; thus,

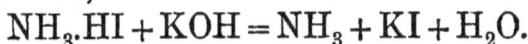

$$NH_3.HI + KOH = NH_3 + KI + H_2O.$$

If the compound NCH_6I, formed by the union of iodomethane $[(CH_3)I]$ and ammonia (NH_3), is properly represented by the formula $NH_3.(CH_3)I$, then we should expect potash to react with it and produce methylamine $[(CH_3)NH_2]$, iodide of potassium, and water ; thus,

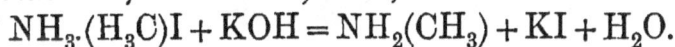

$$NH_3.(H_3C)I + KOH = NH_2(CH_3) + KI + H_2O.$$

Potash does react with the compound formed by the union of iodomethane and ammonia, and the products of the reaction are methylamine, iodide of potassium, and water.

Putting the whole of these reactions together, we are evidently justified in asserting that the reactions of methylamine are expressed by the

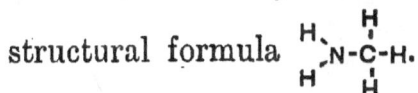

structural formula
$$\begin{smallmatrix} & & \text{H} \\ \text{H}_{\diagdown} & & | \\ & \text{N}{-}\text{C}{-}\text{H}. \\ \text{H}^{\diagup} & & | \\ & & \text{H} \end{smallmatrix}$$

The relations which are shewn to exist between the six compounds methane, chloromethane, methylic alcohol, methylamine, formic aldehyde, and formic acid, by a study of their reactions, are summarised by the structural formulæ which are given to the molecules of these compounds, provided one is acquainted with the language in which these formulæ are expressed. The structural formulæ, in abbreviated form, are these :—

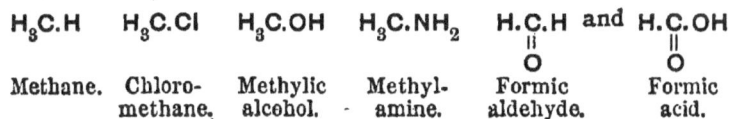

$H_3C.H$	$H_3C.Cl$	$H_3C.OH$	$H_3C.NH_2$	$\underset{\underset{O}{\|\|}}{H.C.H}$ and	$\underset{\underset{O}{\|\|}}{H.C.OH}$
Methane.	Chloromethane.	Methylic alcohol.	Methylamine.	Formic aldehyde.	Formic acid.

Looking at the reactions of the compounds in this series, in a broad and general way, it is noticed that a difference between the functions of the atoms of hydrogen in the molecule of any one of the compounds distinctly begins when methylic alcohol is reached; one, and only one, of the hydrogen atoms in the molecule $H_3C.OH$ can be replaced by the metal potassium. A comparison of some of the reactions of methylic alcohol and formic acid is instructive. The molecules of both of these compounds are represented as containing the group of atoms $-OH$. The hydrogen atom of this group in formic acid can be replaced by a metal when the acid reacts with the hydrooxide of a metal; for instance, $H.CO.OH + KOH = H.CO.OK + H_2O$, and $H.CO.OH + NaOH = H.CO.ONa + H_2O$. But the hydroxides of metals do not react with methylic alcohol; the hydrogen atom of the group $-OH$ in the molecule of this compound can be replaced, it is true, by potassium or sodium (but not by metals in general), but only by causing the alcohol to react with the metal itself in place of using the hydroxide of the metal.

The hydrogen atom in the group $-OH$ in the molecule of formic acid is generally said to be *acidic hydrogen*; because a characteristic reaction of acids, as a class, is, that part or all of their hydrogen can be replaced by metals by causing them to react with hydroxides of metals. If we examine the two formulæ $H_3C.OH$ and $H.CO.OH$ more closely, we see that the carbon atom to which the group $-OH$ is directly attached is in direct union with an atom of oxygen in the molecule of formic acid, but in direct union with three atoms of hydrogen in the molecule of methylic alcohol. Oxygen is a very decided non-metallic element; in many respects, hydrogen is a metal. As we proceed we shall find that when, in a molecule of a compound, an atom of hydrogen is attached, directly or not, to an atom of carbon which also holds directly to itself atoms of a very distinctly non-metallic element, that atom of hydrogen can be replaced by metal in very many cases, by reactions between the compound and metallic hydroxides.

The linking of the markedly non-metallic atoms to the carbon atom impresses acidic functions on the atom of hydrogen which is in union with the carbon atom. The meaning of this statement will be clearer, and it will be possible to give a more definite expression to the statement, when we have become acquainted with the reactions of a number of compounds. Neither of the atoms of hydrogen in the molecule of formic aldehyde (H.CHO) has acidic functions; yet the carbon atom is represented, in the structural formula, as in direct union with an atom of oxygen. But there are two atoms of the non-metallic oxygen in the molecule of formic acid, and only one in that of formic aldehyde; and one of the oxygen atoms in the molecule of the acid is directly linked both to the carbon atom and also to the atom of hydrogen which is acidic in its reactions.

But, as I have said, more knowledge of the reactions of individual compounds, and more

acquaintance with the use of structural formulæ as expressions of reactions, are required
before we can profitably approach, except in
the most general way, questions concerning the
influence, on the functions of this or that atom
in a molecule, of the nature of the other atoms,
and the arrangement of all the atoms, in the
molecule.

Methylic alcohol is known commercially as
Wood Spirit. A very brief account of the manufacture of wood spirit may not be out of place
here. When wood is heated to a high temperature in an iron retort, many different products
are obtained; charcoal and ash remain in the
retort; hydrogen, carbon monoxide and dioxide,
marsh gas, and many other carbon compounds,
pass off as gases; and the liquid portion of the
distillate contains, among other substances,
water, tar, benzene, pyroligneous acid, and
wood spirit. The last named substance is crude
methylic alcohol. The pure alcohol is obtained
by adding lime to the liquid part of the product
of the distillation of wood, then distilling again,
and collecting that portion of the distillate which
is lighter than water, adding a little sulphuric
acid to this distillate and once more distilling,
then adding chloride of calcium which combines
with the methylic alcohol and forms a solid substance; this solid compound is then separated,
and decomposed by distilling it with water, when
the alcohol passes over at a temperature much
below the boiling point of water. Lime is added
to the crude distillate in order that lime salts of

the acids which are present may be formed; when the liquid is then distilled these lime salts remain in the retort. The distillate is shaken with sulphuric acid, in order that salts of the alkaline compounds present may be formed, which salts will remain behind when the next distillation-process is conducted.

These processes, whereby methyl alcohol is obtained from a liquid which contains many other compounds, are examples of the methods generally used for such a purpose. The com-pounds that are not wanted are eliminated, one by one, by causing them to form new combinations, which differ so much in their physical properties from the compound that is to be obtained that they may be separated from that compound by making use of these differ-ences. If the desired compound cannot be separated completely by such processes, it is caused to combine with some other substance to form a new body, which can be purified, and then decomposed with the separation of the wished-for compound.

There are two other compounds derived from marsh gas which demand our attention for a moment. When marsh gas is mixed with three times its volume of chlorine, and the mixture is exposed to sunlight, the main product of the re-action which occurs is a compound that has the composition $CHCl_3$. This compound is called chloroform. A similar combination of carbon, hydrogen, and iodine, which has the composition CHI_3, and is called iodoform, is produced by heating chloroform with hydriodic acid (HI),

F

and also by other reactions. Both chloroform
and iodoform may be prepared from ordinary
alcohol ; the former by heating the alcohol with
bleaching powder, the latter by causing the
alcohol to react with potash and iodine. The
formulæ of the two compounds—$CHCl_3$ and
CHI_3—suggest the names *trichloromethane* and
tri-iodomethane ; and these are the names by
which the compounds are designated in a
systematic method of nomenclature. Chloroform is a liquid with a sweetish taste and a
somewhat spirituous odour ; iodoform is a yellow,
crystalline solid, with a very incisive and per-
sistent smell. Chloroform was discovered by
Liebig in 1831. About 1847 Sir James Simp-
son made use of it as an anæsthetic (that is,
a substance which produces loss of sensibility) in
surgical operations ; and since that time it has
been very much employed for rendering patients
insensible to pain during operations. Iodoform
is much used as an antiseptic (that is, a substance
which prevents putrefaction) in the treatment
of wounds.

CHAPTER VI.

ETHANE AND SOME OF ITS DERIVATIVES.

THE series of compounds to be considered in this
chapter is very similar to the series beginning
with marsh gas and ending with methylamine, that
was noticed in *Chapter V.* Among the gases
which arise from natural petroleum is a hydrocarbon whose composition is expressed by the

formula C_2H_6. This compound is called *ethane*. Treatment of ethane with chlorine produces *chloro-ethane*, C_2H_5Cl; *ethylic alcohol*, C_2H_6O, is formed by the reaction of a boiling solution of caustic potash with chloro-ethane ; ethylic alcohol can be oxidised, first to *acetic aldehyde*, C_2H_4O, and then to *acetic acid*, $C_2H_4O_2$; and *ethylamine*, C_2H_7N, is produced by heating ethylic alcohol with a concentrated watery solution of ammonia. The processes whereby the compounds from chloro-ethane to ethylamine are formed from the hydrocarbon ethane, are evidently very like the processes whereby the corresponding series of compounds, from chloromethane to methylamine, is produced from the hydrocarbon methane ; moreover, the chemical properties of any compound in the first series very closely resemble the chemical properties of the corresponding compound in the second series. Hence we should expect a detailed examination of the reactions of the compounds in the ethane series to shew that the arrangements of the atoms in the molecules of these compounds are similar to the arrangements of the atoms in the molecules of the compounds in the methane series. Assuming the atomic arrangements to be similar in the molecules of the compounds of the two series, we have the following structural formulæ :—

Hydro-carbon.	Chloro-derivative.	Alcohol.	Amine.	Aldehyde.	Acid.
Methane Series—					
H₄C	H₃C.Cl	H₃C.OH	H₃C.NH₂	H.CH | O	H.C.OH || O
Ethane Series—					
H₃C.CH₃	H₃C.CH₂Cl	H₃C.CH₂.OH	H₃C.CH₂.NH₂	H₃C.CH || O	H₃C.C.OH || O

A detailed study of the reactions of ethylic alcohol, ethylamine, acetic aldehyde, and acetic acid confirms the justness of the formulæ assigned to these compounds. It is not necessary to enumerate even the main results of this study; let us rather select a few reactions of the compounds in the ethane series. Ethylic alcohol reacts with the metal potassium, or the metal sodium, to produce the compound C_2H_5KO (or C_2H_5NaO), and hydrogen: only one of the six hydrogen atoms in the molecule of the alcohol can be replaced by potassium or sodium; five atoms of hydrogen always remain in combination with the whole of the carbon and the whole of the oxygen of the molecule C_2H_6O. Acetic aldehyde is very ready to remove oxygen from many compounds of that element, and, by combining with the oxygen so removed, to be changed to acetic acid. Acetic acid is monobasic; that is, it forms only one silver salt, one potassium salt, and one sodium salt: or, to put this reaction into other words, the composition of every compound which is produced by replacing hydrogen in acetic acid by a metal can be expressed by one or other of these formulæ, $H_3C.CO.OM'$, $(H_3C.CO.O)_2M''$, $(H_3C.CO.O)_3M'''$, or $(H_3C.CO.)_4M''''$; where M' is an atom of such a metal as potassium, M'' is an atom of such a metal as copper, M''' is an atom of such a metal as iron, and M'''' is an atom of such a metal as tin. Ethylamine is a colourless liquid, smelling like ammonia: it is very soluble in water, and the solution reacts very similarly to ammonia solution; for instance, precipitates of the hydroxides of iron, copper, lead, tin, and many other metals, are formed when a solution of ethylamine is added to solutions of salts of these metals, just as the same hydroxides are precipitated from the same salts by using a solution of ammonia. As methylic alcohol ($H_3C.OH$) reacts with hydrochloric acid gas, in the presence of some substance which readily combines with water, to produce chloromethane ($H_3C.Cl$), so ethylic alcohol ($H_3C.CH_2.OH$) reacts under similar conditions to produce chloro-ethane ($H_3C.CH_2Cl$).

Let us look a little more closely at the structural formulæ which express the reactions of methylic and ethylic alcohols. Let us write these formulæ in full, thus :—

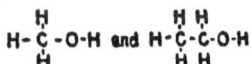

$$\begin{array}{ccc} & H & & H \ H \\ H- & \!\!C\!\!-O\!\!-H & \text{and} & H-\!\!C\!\!-\!\!C\!\!-O\!\!-H \\ & H & & H \ H \end{array}$$

In each molecule an atom of carbon is represented in direct union with an atom of oxygen which is directly linked to an atom of hydrogen, and the same atom of carbon is represented in direct union with two atoms of hydrogen ; in the molecule of methylic alcohol this carbon atom is shewn as directly joined to a third atom of hydrogen, while in the molecule of ethylic alcohol it is shewn as directly joined to the group of atoms CH_3. Looking at the formulæ in this way, we may express them thus :—$H.CH_2OH$ and $H_3C.CH_2OH$. Both molecules *contain the atomic group CH_2OH.* Now it is customary to say that this group of atoms is the characteristic group of the molecules of a great many compounds all of which are alcohols. All these alcohols have certain common properties; for instance, they all oxidise to aldehydes—each aldehyde containing the same number of atoms of carbon and oxygen as the alcohol contains, and two atoms of hydrogen less than the alcohol contains—and these aldehydes then oxidise to acids, which contain the same number of atoms of carbon as the parent alcohols, but one atom of oxygen more, and two atoms of hydrogen less, than the parent alcohols. The fact that these alcohols have many chemical properties in common, is connected with the atomic structure of their molecules by saying that these molecules all contain the group CH_2OH.

I ask the reader to pay especial attention to this statement, *the molecules of all the alcohols of a certain class contain the atomic group CH_2OH*, and to consider it carefully in the light of the reactions of methylic and ethylic alcohols. The statement summarises many reactions, and suggests other reactions; but it does these things only to the person who understands the special language of the molecular and atomic theory.

As another illustration of the use of this very convenient phrase, *certain molecules contain a common atomic group*, let us look a little closely at the structural formulæ which summarise the reactions of formic and acetic aldehydes. These

formulæ are $H-\underset{O}{\overset{}{C}}-H$ and $H-\underset{H\ O}{\overset{H}{C}}-\overset{}{C}-H$; or, more shortly,

H.CHO and H₃C.CHO.

The atomic group common to the two molecules is HCO, or more fully, $\overset{}{\underset{O}{C}}{}^{-H}$. There are many other aldehydes which behave similarly to

methylic and ethylic aldehydes, under similar
conditions : all are easily oxidised to acids by
combining with an atom of oxygen ; all remove
oxygen from many other compounds, and are,
therefore, said to act as reducing (or deoxidising)
agents. The reactions of these aldehydes are
summed up in the one structural formula,
R.CHO ; where R represents an atom of hydrogen in the case of methylic aldehyde (which is the
first aldehyde of the series), or a group of atoms
of carbon and hydrogen in all cases except that
of methylic aldehyde. Readiness to remove
oxygen from compounds which do not insist
on retaining all their oxygen is a reaction common
to certain compounds ; the molecules of these
compounds are thought of as containing an
arrangement of carbon, hydrogen, and oxygen
atoms which is presented in intelligible form by

the symbol $\overset{C-H}{\underset{O}{}}$. Now turn to the structural for-

mulæ of formic and acetic acids, $H.\overset{C.OH}{\underset{O}{}}$ and $H_3C.\overset{C}{\underset{O}{}}OH$;

and compare these with the structural formulæ of

the two aldehydes $H.\overset{CH}{\underset{O}{}}$ and $H_3C.\overset{CH}{\underset{O}{}}$. The molecule of

formic acid is represented as containing the
group H.CO ; whereas the molecule of acetic acid
is not represented as containing this group. But
H.C.O is the characteristic aldehydic group: therefore, if the formula given to formic acid is a truthful
representation (in its own language) of the reactions of that acid, we should expect formic acid
to act as a remover of oxygen from compounds
which are fairly easily deprived of their oxygen,
just as aldehydes act as removers of oxygen ; and

we should not expect acetic acid to act as an energetic deoxidising agent. These expectations are fulfilled when the reactions of the two acids with fairly easily deoxidised compounds are examined. For instance, a cold solution of permanganate of potassium (Condy's fluid) is deoxidised by formic acid, but not by acetic acid; gold, platinum, and palladium are precipitated from solutions of their salts by formic acid (that is, the formic acid decomposes the salts of these metals and combines with the oxygen in them), but not by acetic acid; and an alkaline solution of a salt of copper is reduced by formic acid, but not by acetic acid.

This comparison of the structural formulæ of formic and acetic acids, and of these formulæ with those of two aldehydes, shews how suggestive of reaction these formulæ are, and, therefore, how much they may help to advance the study of the connexions between composition and properties the elucidation of which connexions is the goal of chemical inquiry.

When ethylic alcohol is heated with nearly double its weight of sulphuric acid, in a flask connected with a condensing apparatus, and more ethylic alcohol is allowed to drop into the hot mixture, a colourless, mobile, very inflammable, liquid collects in the receiver. This liquid is *ether*.

When a mixture of ethylic alcohol and acetic acid is heated, and the gaseous product of the reaction is condensed, a colourless, very fragrant, liquid is obtained. This liquid is *ethylic acetate.*

What relations, of properties, and of composition, exist between these two compounds, ether and ethylic acetate, and ethylic alcohol from which they are prepared? The composition of ether is given by the formula $C_4H_{10}O$, and that of ethylic acetate by the formula $C_4H_8O_2$. These formulæ do not suggest any definite relations between the two compounds and ethylic alcohol (C_2H_6O). Let us look at the reactions of the two compounds.

When ether is treated with hydrochloric acid gas, chloro-ethane (C_2H_5Cl) is formed; if the quantity of ether which reacts, and the quantity of chloro-ethane which is produced, are determined, the interaction is found to be expressed by the chemical equation $C_4H_{10}O + 2HCl = 2C_2H_5Cl + H_2O$. The atom of oxygen is removed from the molecule $C_4H_{10}O$, and two atoms of chlorine are put in its place; but instead of one molecule, having the composition $C_4H_{10}Cl_2$, being formed, two molecules are produced each having the composition C_2H_5Cl. Now we found before (p. 67) that methylic alcohol and hydrochloric acid react to produce chloromethane, the reaction being represented in structural formulæ thus,—

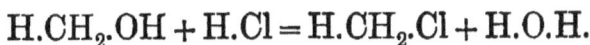

$$H.CH_2.OH + H.Cl = H.CH_2.Cl + H.O.H.$$

We also noticed (p. 73) that formic aldehyde reacts with phosphorus pentachloride, which is a compound that acts like hydrochloric acid but more energetically than that acid, to form CH_2Cl_2; thus,—

$$\overset{\text{H.C.H}}{\underset{\text{O}}{\overset{\|}{}}} + PCl_5 = \overset{\text{H.C.H}}{\underset{\text{Cl}_2}{\overset{\|}{}}} + POCl_3 \text{ (oxychloride of phosphorus).}$$

Looked at in the light of these reactions, the interaction of hydrochloric acid and ether suggests that the oxygen atom in the molecule of ether is in direct union with a carbon atom, or with carbon atoms, and is not linked to an atom of hydrogen which is again linked to an atom of carbon.

Let us turn to another reaction of ether. Ether and water act on one another to produce ethylic alcohol; the change is presented in an equation in this way,—

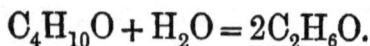

$$C_4H_{10}O + H_2O = 2C_2H_6O.$$

Remembering that the reaction of ether with hydrochloric acid suggests the direct linkage of the oxygen atom in the molecule of ether to carbon, and recalling the structural formula of ethylic alcohol, and the close relationship which evidently exists between that alcohol and ether, we try the following scheme as a possible representation of the change that occurs when ether reacts with water to form ethylic alcohol:—

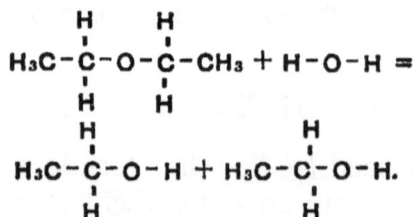

$$
\begin{array}{ccc}
& H & H \\
& | & | \\
H_3C- & C-O-C & -CH_3 + H-O-H = \\
& | & | \\
& H & H
\end{array}
$$

$$
\begin{array}{cc}
H & H \\
| & | \\
H_3C-C-O-H + & H_3C-C-O-H. \\
| & | \\
H & H
\end{array}
$$

If this presentation of the changes is a correct translation of the process into the language we are trying to learn, then ether is an oxide of the atomic group $H_3C.CH_2$; and ethylic alcohol is a compound of that group with oxygen and hydrogen, that is to say, it is an hydroxide of the same atomic group. The name *ethyl* is given to the group of atoms $H_3C.CH_2$ which is supposed to exist in the molecules of ether and ethylic alcohol; supposed to exist in these molecules, let it be remembered, because of the reactions of the two compounds. Let us employ the symbol *Et* to represent the group ethyl; then the formulæ of ether and ethylic alcohol are represented by these symbols:—Et.O.Et and Et.OH. We may call these two compounds *ethyl oxide* and *ethyl hydroxide*, when we wish to emphasise the conception we have formed of the relations between them. By using these names we imply that the compounds are similar to such inorganic compounds as potassium oxide and potassium hydroxide; K.O.K and K.OH. Attention might be drawn to other reactions of ether, but I think it will be sufficient to say that these reactions are quite in keeping with the representation of this compound as ethylic oxide.

But what are the relations of ethylic acetate, $C_4H_8O_2$, to ethylic alcohol, and to acetic acid, the two compounds which react to produce the acetate? One of the two compounds which react to produce ethylic acetate is an acid, and we have learned that the other compound, ethylic alcohol, is like potassium hydroxide in its reactions. Now what is the ordinary reaction between an acid and a metallic hydroxide? Sulphuric acid and sodium hydroxide interact to produce sodium sulphate, which is a salt, and water; nitric acid and potassium hydroxide interact to form potassium nitrate,

which is a salt, and water; hydrochloric acid
and barium hydroxide interact to produce the
salt barium chloride, and water; acetic acid and
lead hydroxide interact to produce the salt lead
acetate, and water. The ordinary products of
the reaction between an acid and a metallic
hydroxide are a salt, formed of the metal of the
hydroxide and the radicle of the acid, and
water. It is probable then that acetic acid
and ethyl hydroxide will react to produce a
salt and water; and we should expect the salt
to be composed of the radicle ethyl, which plays
the part of a metal in the hydroxide ethylic
alcohol, and the radicle of acetic acid. But
what is the composition of the radicle of acetic
acid? This question is answered by tabulating
the compositions of some salts of acetic acid—
these salts are called *acetates*—and comparing
these compositions with that of acetic acid. The
following table presents data sufficient for our
purpose.

Acetic acid. $C_2H_4O_2$.
Acetates.

$C_2H_3O_2Na$.	$C_2H_3O_2K$.	$(C_2H_3O_2)_2Ba$.	$(C_2H_3O_2)_2Pb$.
Sodium acetate.	Potassium acetate.	Barium acetate.	Lead acetate.

One, and only one, of the four atoms of hydrogen in the molecule $C_2H_4O_2$ is replaced by metal
when acetates are formed : the radicle of acetic
acid has the composition C_2H_3O. We should,
then, expect the reaction between acetic acid and
ethyl hydroxide to result in the formation of a
salt, composed of the radicle or atomic group ethyl
(C_2H_5) in combination with the radicle of acetic
acid, C_2H_3O. Moreover, as ethyl hydroxide

(C_2H_5OH) is similar to potassium hydroxide (KOH), we should expect the salt which is formed, according to the hypothesis that is guiding us, by the reaction of acetic acid with ethyl hydroxide, to have a composition analogous to the composition of potassium acetate; that is to say, we should expect the composition of the salt to be expressed by the formula $C_2H_3O_2(C_2H_5)$. Now the formula which expresses what we may call the empirical composition of the compound in question—that is, the composition without taking into account what may be the structure of the molecule of the compound—is $C_4H_8O_2$. The symbol C occurs four times, the symbol H occurs eight times, and the symbol O occurs twice, in the formula $C_2H_3O_2(C_2H_5)$; in other words, $C_2H_3O_2(C_2H_5) = C_4H_8O_2$. The product of the reaction between acetic acid and ethylic alcohol (ethyl hydroxide) is then, probably, the salt ethyl acetate, $C_2H_3O_2(C_2H_5)$. A closer examination of the reactions of this substance shows that it belongs to the class of salts; and a quantitative study of the change that occurs when acetic acid and ethylic alcohol react proves that this change is represented, correctly, by the equation—

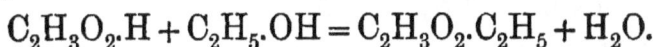

$$C_2H_3O_2.H + C_2H_5.OH = C_2H_3O_2.C_2H_5 + H_2O.$$

Alcohols are hydroxides of certain atomic groups, or radicles; and ethers are oxides of the same atomic groups. The salts that are formed by the replacement of the acidic hydrogen of acids by these atomic groups are called *ethereal salts*. Ethylic acetate is the first example we have had of an ethereal salt.

The reaction which occurs when an ethereal salt is heated with a solution of caustic potash or caustic soda, is one of considerable importance. Let us consider this reaction for a moment, and let us first of all see what change takes place when a metallic salt reacts with a solution of caustic potash. Let lead acetate be taken as the example of a metallic salt. The reaction between lead acetate and caustic potash solution results in the formation of lead hydroxide and potassium acetate ; the change is expressed thus in an equation—

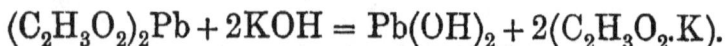

$$(C_2H_3O_2)_2Pb + 2KOH = Pb(OH)_2 + 2(C_2H_3O_2.K).$$

This is a special case of the change that generally occurs when caustic potash, or soda, reacts with a metallic salt; what may be called the normal products of this reaction are, a hydroxide of the metal of the salt, and a salt of potassium (or sodium) composed of that metal united with the acidic radicle of the original salt. Now when the ethereal salt ethylic acetate reacts with a solution of caustic potash (if it does react), we should expect the products to be, hydroxide of ethyl and potassium acetate ; that is, we should expect the change to be expressed by this equation—

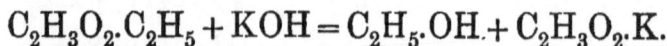

$$C_2H_3O_2.C_2H_5 + KOH = C_2H_5.OH + C_2H_3O_2.K.$$

And this is indeed what happens. The reaction proceeds slowly : in order to change the whole of the ethylic acetate, considerably more potash, in solution, must be used than is expressed in the equation ; the reacting compounds must be kept in contact, at a high temperature, for a long time ;

and the ethylic alcohol that is produced must be removed, by distillation, from time to time as the change proceeds. We shall have other examples of the reaction between ethereal salts and caustic potash, or soda, especially when we are considering the processes of making soaps (in chapters VII. and VIII.). This reaction produces alcohols, and potassium (or sodium) salts of the acids of the ethereal salts employed.

In this chapter, and in chapter V. we have been concerned with two similar series of compounds of carbon. Each series begins with a *hydrocarbon* whose composition is given by the formula C_nH_{2n+2}; by the action of chlorine on the hydrocarbon a *chloro-derivative* is obtained which has the composition $C_nH_{2n+1}Cl$; this chloro-derivative of the hydrocarbon reacts slowly with a watery solution of caustic potash to form an *alcohol*, $C_nH_{2n+1}.OH$; the alcohol reacts with ammonia, or with ammonium chloride, to produce an *amine*, $C_nH_{2n+1}.NH_2$; the alcohol also reacts with oxidising agents to yield, first an *aldehyde*, $C_nH_{2n}O$, and then an *acid*, $C_nH_{2n}O_2$. In the second series, we noticed, also, that the alcohol reacts with sulphuric acid to produce an *ether*, $(C_nH_{2n+1})_2O$; and that the acid and the alcohol interact to form an *ethereal salt*, $C_nH_{2n-1}O_2.C_nH_{2n+1}$. It is possible, by appropriate reactions, to obtain an ether from the alcohol of the first series, and also an ethylic salt of the acid of that series; the compositions of these two compounds are given by the formulæ, $(CH_3)_2O$ for methylic ether, and $CHO_2.C_2H_5$ for ethylic formate. The relations of composition of the members of the two series are perhaps more fully expressed by using the following general formulæ :—

General Formula.	Examples.
HYDROCARBON.—C_nH_{2n+2}	CH_4 (methane) and C_2H_6 (ethane).
CHLORO-DERIVATIVE.—$C_nH_{2n+1}Cl$	CH_3Cl (chloromethane) and C_2H_5Cl (chloro ethane).
AMINE.—$C_nH_{2n+1}.NH_2$	CH_3NH_2 (methylamine) and $C_2H_5NH_2$ (ethylamine).
ALCOHOL.—$C_nH_{2n+1}.CH_2OH$	$H.CH_2OH$ (methylic alcohol) and $H_3C.CH_2OH$ ethylic (alcohol).
ALDEHYDE.—$C_nH_{2n+1}.CHO$	$H.CHO$ (formic aldehyde) and $H_3C.CHO$ (acetic aldehyde).
ACID.—$C_nH_{2n+1}.CO.OH$	$H.CO.OH$ (formic acid) and $H_3C.CO.OH$ (acetic acid).
ETHEREAL SALT.— $C_nH_{2n+1}.CO.O(C_mH_{2m+1})$	$H.CO.O(C_2H_5)$ (ethyl formate) and $H_3C.CO.O(C_2H_5)$ (ethyl acetate).
ETHER.—$(C_nH_{2n+1})_2O$	$(CH_3)_2O$ (methyl ether) and $(C_2H_5)_2O$ (ethyl ether).

Note.—In the general formula $n = 0$ in the cases of methylic alcohol, formic aldehyde, formic acid, and ethyl formate.

The examination of the compounds in these two series has brought with it many examples of the use of the conception of the *compound radicle,* or atomic group which holds together throughout many reactions and impresses certain common properties on all the molecules of which it forms a part. The study of the two series of compounds has also furnished illustrations of the meaning of the term *substitution ;* we have had cases of the substitution of one atom by another, for instance in the passage from CH_4 to CH_3Cl and from C_2H_6 to C_2H_5Cl; we have had cases of the substitution of an atom by an atomic group, for instance in the passage from CH_3Cl, or C_2H_5Cl, to $CH_3.OH$, or $C_2H_5.OH$; and we have had cases of the substitution of one atomic group by another, for instance in the passage from $CH_3.OH$, or $C_2H_5.OH$, to $CH_3.NH_2$, or $C_2H_5.NH_2$. The attempt made in this, and the previous

chapter to connect the reactions of certain compounds of carbon with the compositions of these
compounds has shewn, I hope, how impossible it
is to elucidate the relations of reactions to composition without the help of some definite theory
of the structure of matter, and the aid of not
a few subsidiary hypotheses and conventions
whereby the application of the theory is made
practicable to chemical phenomena.

Finally, the general description of the features
of the chemical changes that occur between
compounds of carbon, given in Chapter II. (pp.
31-34), has, I think, been justified by what we
have learned of the changes in the two series of
compounds derived, respectively, from marsh
gas and ethane. In some respects, the hydrocarbons may be regarded as representatives
among carbon compounds of the metals of inorganic chemistry. When one desires to form
a salt of a metal, it is sufficient to dissolve the
metal in the appropriate acid and to evaporate
the solution. But in order to make a salt from
a hydrocarbon, it is necessary to prepare a chloro-
derivative of the hydrocarbon, from this to pass
to the alcohol, and then, through the aldehyde, to
the acid ; having obtained the acid and the
alcohol, it is still necessary to cause these to
react in order to obtain the wished-for salt. To
prepare an oxide of a metal, it is generally sufficient to burn the metal in oxygen ; but between
the hydrocarbon and the oxide, which is an
ether, how many stages there are ! The alcohol
is obtained after two operations, and a third process of change is required before the oxide appears.

G

In Chapter VIII. I shall give a short account of the technical applications of some of the compounds whose reactions and compositions have formed the subject-matter of Chapters V. and VI.

CHAPTER VII.

ETHYLENE, GLYCERIN, AND TARTARIC ACID.

ONE of the many hydrocarbons present in ordinary coal-gas is called *ethylene*, and has the composition C_2H_4. This compound is generally prepared in the laboratory by heating ethylic alcohol with about four times its volume of concentrated sulphuric acid; the principal reaction that occurs is represented by the equation $C_2H_6O + H_2SO_4 = H_2SO_4.H_2O + C_2H_4$. Ethylene is a colourless gas, slightly lighter than air; when cooled a little below the freezing point of water, and subjected to a pressure of about 44 atmospheres (about 650 lbs. on the square inch), the gas changes to a colourless liquid which boils at *minus* 103°C. [*minus* 153°F.].

Ethylene is easily burnt in the presence of air. When the gas is mixed with a little air and burnt, the main products of the combustion are carbon dioxide, water, and carbon. Under these conditions the gas burns with a luminous flame; and the luminosity is caused, chiefly, by the minute solid particles of carbon becoming red hot and radiating light. If ethylene is mixed with somewhat more than three times its own volume of oxygen (equal to more than fifteen volumes of air) and the mixture is burnt, only carbon dioxide and water are formed, and the flame is almost non-luminous. Part of the luminosity of an ordinary flame of coal-gas is due to the limited combustion of the ethylene which the gas contains. When coal-gas issues from a burner, admixture with air occurs at the edges of the stream of gas ; and when a lighted taper is brought near the burner chemical change begins where the air and the gas are in contact ; the combustion of the ethylene, and the other hydrocarbons in the gas, is, therefore, not complete ; solid particles of carbon are produced, and these are heated to so high a temperature that they give out much light. If coal-gas is mixed with a large quantity of air and the mixture is ignited, the combustion of the ethylene, and other hydrocarbons, is practically complete, and all the products are gaseous. The flame of such a mixture is extremely hot but almost free from light. The Bunsen-lamp, which is employed in all laboratories for getting high temperatures, is an extremely simple contrivance for securing the admixture with coal-gas of sufficient air to insure the complete combustion of the carbon compounds in the gas. The apparatus consists of an ordinary gas-burner over which is fitted an iron tube, about 3 inches long and $\frac{3}{8}$ inch diameter, having a couple of holes (about $\frac{1}{4}$ inch diameter) pierced near its lower end. Air flows in through the holes, and the mixture of gas and air is ignited at the upper end of the tube. The principle of the Bunsen-lamp is applied to the construction of gas-stoves. Sometimes the stove

consists of several small upright tubes, into each
of which gas and air are admitted at the bottom,
and the mixture is burnt at the top. Sometimes
there is one rather wide tube pierced with many
small holes or slits ; this tube is attached to the
gas-supply, and there is a large opening, near
where the gas enters, to admit air ; the wide
tube becomes filled with a mixture of gas and
air, and this mixture is burnt as it issues from
the small holes or slits. If too much air is
mixed with the gas in a Bunsen-burner, or a
gas-stove, the temperature of the flame is de-
creased by the superfluous air ; if too little air
is admitted the combustion of the carbon compounds of the gas is not complete, and the flame
is smoky.

We have learnt, in chapters V. and VI., that
both marsh gas (CH_4) and ethane (C_2H_6) react
with chlorine to produce compounds containing
chlorine in place of part of the hydrogen of the
marsh gas or the ethane. The reaction of ethylene (C_2H_4) with chlorine is not similar to the
reactions of the two hydrocarbons we have considered. Ethylene combines with chlorine slowly,
in diffused daylight, to form an oily liquid, with
a sweetish odour, called *ethylene chloride.* The reaction is expressed thus in an equation ;—C_2H_4
$+ 2Cl = C_2H_4Cl_2$. The chlorine does not remove
hydrogen from the molecule of ethylene, as it
does from the molecules of marsh gas and
ethane ; but the chlorine adds itself to the
molecule of ethylene. Bromine reacts with ethylene in a way exactly similar to that wherein
chlorine reacts ; a compound of ethylene and

bromine is formed, called *ethylene bromide*, and having the composition $C_2H_4Br_2$. It is customary to distinguish such compounds as CH_3Cl and C_2H_5Cl (chloromethane and chloro-ethane), that are produced by the substitution of an atom (or atoms) in one molecule by another atom (or by other atoms), from such compounds as $C_2H_4Cl_2$ and $C_2H_4Br_2$ (ethylene chloride and ethylene bromide) that are produced by the addition of an atom, or atoms, to a molecule of a compound. Compounds of the former class are called *substitution-compounds*, and compounds of the latter class are named *additive compounds*. It is also customary to speak of compounds, like marsh gas and ethane, which always react to form substitution-compounds as *saturated* ; and compounds, like ethylene, which react to produce additive compounds as *unsaturated*. These four terms are convenient aids towards remembering the general reactions of carbon compounds. A hydrocarbon is being examined : it is found to combine readily with chlorine, and to form a compound composed of chlorine added on to the whole of the molecule of the hydrocarbon ; the hydrocarbon is placed in the class of unsaturated compounds ; and many of its other reactions are known, because all unsaturated compounds exhibit certain common reactions.

Ethylene chloride $(C_2H_4Cl_2)$ and ethylene bromide $(C_2H_4Br_2)$ are oily liquids. In the year 1795, four Dutch chemists — named Deiman, Pæts van Troostuyk, Bondt, and Lauwerenburgh—discovered a new gas, produced by the action of oil of vitriol on common alcohol ; they found that the gas was composed of carbon and hydrogen only ; and they noticed that an oily liquid was formed when this gas was brought into contact with the substance now called chlorine. The Dutch chemists gave the name *gaz huileux* (oily gas) to the new compound, a name that was afterwards changed to *gaz oléfiant* (oil-forming gas). The oily compound formed by the union of this gas with chlorine was known for a long time by the names *Dutch liquid* and *oil of the Dutchmen*; these names have disappeared, but the compound which is produced by the reaction of oil of vitriol and alcohol is still commonly spoken of as *olefiant gas.*

Ethylene chloride and ethylene bromide react gradually with caustic potash solution to form a compound which belongs to the class of alcohols. The composition of this compound is $C_2H_4(OH)_2$; because of its sweet taste it is called *glycol.* (The termination *-ol* is common to the names of all compounds which are alcohols). The reactions of ethylene are expressed, in terms of the atomic and molecular theory and the hypothesis of

atom-linking, by the structural formula
$$
\begin{array}{c} H \quad H \\ | \quad | \\ C = C \\ | \quad | \\ H \quad H \end{array}
$$

or more shortly $H_2C = CH_2$. The reaction between ethylene and chlorine, and that between ethylene chloride and caustic potash, are presented by the following schemes :—

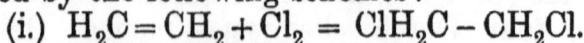

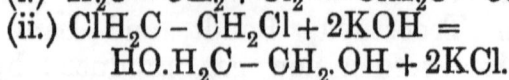

(i.) $H_2C = CH_2 + Cl_2 = ClH_2C - CH_2Cl.$

(ii.) $ClH_2C - CH_2Cl + 2KOH =$
$$HO.H_2C - CH_2.OH + 2KCl.$$

I do not intend to discuss the reactions of

the alcohol, $C_2H_4(OH)_2$, derived from ethylene; I merely wish to note its existence, and the method of its formation.

If the compositions of the two hydrocarbons methane and ethane, CH_4 and C_2H_6, are compared, it is seen that both are expressed by the formula C_nH_{2n+2}; that is to say, the number of hydrogen atoms in either molecule is equal to twice the number of atoms of carbon in the molecule *plus* two. These compounds are the first and second members respectively of a series of hydrocarbons all of which have the common composition C_nH_{2n+2}. The series is known as the *paraffins*, because of their indifference to the action of chemical reagents (*parum affinis* = not much affinity). The third member of the series of paraffins ($n = 3$) is C_3H_8. This hydrocarbon, called *propane*—the names of all the paraffins end in -*ane*—is found in the gases given off from certain petroleum springs in America. Propane reacts with bromine to form various substitution-compounds; one of these, called *tribromopropane*, has the composition $C_3H_5Br_3$; when this compound is boiled with caustic potash the following reaction occurs:—

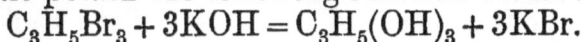

$$C_3H_5Br_3 + 3KOH = C_3H_5(OH)_3 + 3KBr.$$

The compound $C_3H_5(OH)_3$ is an alcohol: it is generally known by the name *glycerin*; but it is better called *glycerol*, because the syllable -*ol* is the characteristic termination of the names of those compounds which are alcohols. Just as the atomic group C_2H_5, that is present in the molecules of ethylic alcohol, ethylic ether, and ethylic acetate, is called *ethyl*, so the atomic

group C_3H_5, is called *glyceryl*. As we spoke of
ethylic alcohol by the name *ethyl hydroxide,* so we
may speak of glycerin as *glyceryl hydroxide.* It
is to be noticed that while a molecule of ethyl
hydroxide is composed of the atomic group
ethyl combined with one atom of oxygen and
one atom of hydrogen, a molecule of glyceryl
hydroxide is composed of the atomic group
glyceryl combined with three atoms of oxygen
and three atoms of hydrogen. As ethyl hydroxide
($C_2H_5.OH$) is analogous, in composition and in
many reactions, to potassium hydroxide (KOH)
and sodium hydroxide (NaOH), so glyceryl hydroxide is similar, in composition and in some
reactions, to ferric hyroxide [$Fe(OH)_3$], and
aluminium hydroxide [$Al(OH)_3$].

When tallow, linseed oil, palm oil, whale oil,
or any one of several other oils, is boiled with
caustic soda, or potash, or with slaked lime,
glycerin is formed along with a compound which
is a soap. The main constituents of fats and
oils are compounds of glyceryl with the radicles
of certain acids that are often classed together
under the name *fatty acids.* These compounds
are ethereal salts ; they belong to the same class
of compounds as ethylic acetate. Turn back for
a moment to the reaction (considered on p. 91)
between ethylic alcohol and acetic acid whereby
ethylic acetate and water are formed, and to the
reaction between ethylic acetate and caustic
potash (considered on p. 94) whereby ethylic
alcohol and potassium acetate are produced.
These reactions find their expressions in the
equations—

(i.) $C_2H_5.OH + C_2H_3O_2.H = C_2H_3O_2.C_2H_5 + H_2O$,

Ethylic alcohol.   Acetic acid.   Ethylic acetate.

(ii.) $C_2H_3O_2.C_2H_5 + KOH = C_2H_5.OH + C_2H_3O_2.K$.

Ethylic acetate.   Potash.   Ethylic alcohol.   Potassium acetate.

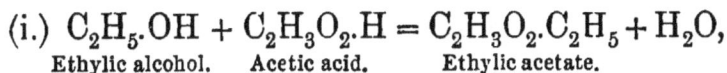

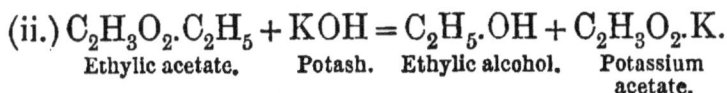

Remembering that the radicle ethyl (C_2H_5), which combines with one atomic group OH, replaces one atom of hydrogen in an acid to form an ethylic salt, we might expect that the radicle glyceryl (C_3H_5), which combines with three atomic groups OH, would replace three atoms of hydrogen in an acid to form a glyceryl salt. Examination of the glyceryl salts shews that this expectation is correct. *Glyceryl acetate* is C_3H_5 $(C_2H_3O_2)_3$, *glyceryl nitrate* is C_3H_5 $(NO_3)_3$, *glyceryl chloride* is $C_3H_5Cl_3$, and so on. These salts may be produced by causing glycerin [$C_3H_5(OH)_3$] to react with the appropriate acid : acetic, nitric, and hydrochloric acids are mono-basic, in other words, a molecule of any one of these acids contains only one atom of hydrogen which can be replaced by metals or by such atomic groups as ethyl or glyceryl; therefore, when glycerin reacts with one of these acids to form a glyceryl salt, three molecules of the acid react with one molecule of glycerin. The reaction, in the case of nitric acid for instance, is expressed in an equation in this way—

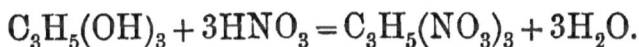

$$C_3H_5(OH)_3 + 3HNO_3 = C_3H_5(NO_3)_3 + 3H_2O.$$

The chief constituent of tallow is *glyceryl stearate*, of linseed oil *glyceryl oleate*, and of palm oil *glyceryl palmitate*. Stearic, oleïc, and palmitic

acids are monobasic acids; their compositions
are given by the formulae—

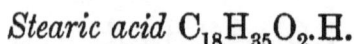

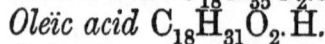

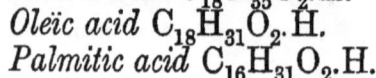

Stearic acid $C_{18}H_{35}O_2.H.$
Oleïc acid $C_{18}H_{31}O_2.H.$
Palmitic acid $C_{16}H_{31}O_2.H.$

And the glyceryl salts of these acids are—

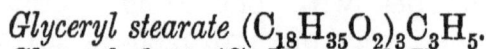

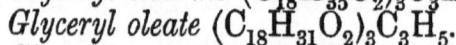

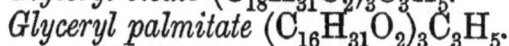

Glyceryl stearate $(C_{18}H_{35}O_2)_3C_3H_5.$
Glyceryl oleate $(C_{18}H_{31}O_2)_3C_3H_5.$
Glyceryl palmitate $(C_{16}H_{31}O_2)_3C_3H_5.$

The reaction that occurs when one of these salts
is boiled with caustic potash solution is exactly
similar to that which occurs when ethylic
acetate is boiled with caustic potash solution.
Selecting glyceryl stearate, the main constituent
of tallow, we express the reaction by the following equation :—

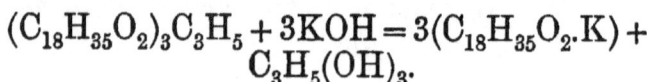

$$(C_{18}H_{35}O_2)_3C_3H_5 + 3KOH = 3(C_{18}H_{35}O_2.K) + C_3H_5(OH)_3.$$

The equations which express the reactions of
caustic potash with the other glyceryl salts are
similar to this. If caustic soda is used, glycerin
and the sodium salt of the acid of the glyceryl
salt are formed. The sodium salts, and the
potassium salts, of stearic, palmitic, and oleïc
acids, and of certain other acids analogous to
these, are soaps: the sodium salts are called
hard soaps, and the potassium salts are called
soft soaps.

The process that takes place when the glyceryl
salt of a fatty acid reacts with potash, or soda,
to form glycerin and a potassium, or sodium, salt

of the fatty acid, which salt is a soap, is called *saponification.* It is customary to apply this term to all cases of the reaction of potash, or soda (or other alkali), with an ethereal salt; for instance, the reaction between ethylic acetate and potash is a particular case of saponification, although potassium acetate has not the physical properties of a soap.

Among the many acids that are found in the juices of fruits there is one which demands our attention at this point in our attempt to follow some of the wanderings of carbon. That compound is *tartaric acid.* This acid, or one of its salts, is found in grape-juice, potatoes, Jerusalem artichokes, the berries of the mountain-ash, pineapples, the juice of beetroots, and in many other plants. As the juice of grapes ferments, a reddish solid is deposited on the sides and bottoms of the vessels which contain the fermenting liquid. This solid matter consists chiefly of potassium tartrate mixed with colouring matters from the grape-juice. The crude potassium tartrate is dissolved in water, the solution is filtered, evaporated, and allowed to cool: the crystals of potassium tartrate that separate from the cold liquid are dissolved in boiling water, and chalk is added to this liquid; a reaction occurs whereby calcium tartrate and potassium carbonate are produced, the latter dissolves in the water, and most of the calcium tartrate remains undissolved as a white solid substance. The calcium tartrate is washed, and then decomposed by the proper quantity of dilute sulphuric acid; tartaric acid and calcium sulphate are formed; the acid dissolves and the calcium sulphate is insoluble. The solution is filtered, and evaporated ; and crystals of tartaric acid form in the cooling liquid. Tartaric acid forms white crystals, the composition of which is expressed by the formula $C_4H_6O_6$.

The property of tartaric acid to which I ask the special attention of the reader is a physical rather than a chemical property. The property is this :—a solution of tartaric acid in water *rotates the plane of polarisation of a ray of light* in the same direction as that in which the hands ot a watch turn when the watch is looked at in the ordinary way. What is *the plane of polarisation* of a ray of light ? And what is meant by the *rotation* of this plane ? Common light is supposed to consist of vibrations of a something called " ether " ; some of these vibrations of the ether are thought of as taking place in one plane, and some in a plane at right angles to the first. Under certain conditions, for instance when a ray of light is reflected from a bright surface, one set of vibrations is stopped. Such a ray of light is said to be *plane-polarised* ; all the vibrations are now occurring at right angles to a certain plane, which is called *the plane of polarisation.* Some substances do not allow a ray of plane-polarised light to pass through them without modifying the vibrations of the ray. The general result of these modifications is this :— before the ray of plane-polarised light enters the substance that modifies it we may suppose all the vibrations to be taking place in a direction at right angles to the surface of the earth ; when the ray leaves the modifying substance it has

been separated into two rays, the vibrations in
which take place in different planes, and as
these rays move with different velocities the
effect of the modifying substance is equivalent
to the turning round, or rotation, of the plan
of polarisation of the ray, either in the same
direction as that in which the hands of a watch
move, or in the direction opposite to that in
which the hands of a watch move, when the
watch is looked at in the ordinary way. Those
substances which produce a *rotation of the plane
of polarisation* of a ray of light are said to be
optically active; all other substances are said to
be *optically inactive*. Substances which cause the
rotation of the plane of polarisation in the same
direction as that in which the hands of a watch
move—or, we may say, substances which cause
watch-like rotation of the plane of polarisation—
are said to be *dextro-rotatory*; and substances
which cause the rotation of the plane of polarisation in a direction opposite to that in which the
hands of a watch move are called *laevo-rotatory*
bodies.

Tartaric acid, in solution, is an optically
active compound. Tartaric acid, in solution,
is an optically inactive compound. Tartaric
acid, in solution, is a dextro-rotatory compound.
Tartaric acid, in solution, is a laevo-rotatory
compound. These four statements are true. In
other words there are four tartaric acids. The
molecule of each of these acids has the composition given by the formula $C_4H_6O_6$. Evidently we must inquire somewhat closely into
the reactions of the tartaric acids in order that

we may connect the differences in their optical activities with differences of molecular structure. Ethylene (C_2H_4) combines readily with bromine (see p. 101) to form ethylene bromide ($C_2H_4Br_2$); by boiling this compound with a solution of potassium cyanide (KCN) in alcohol, a compound called *ethylene cyanide* is produced; when this compound, which has the composition $C_2H_4(CN)_2$, is boiled with water it is slowly changed to ammonia, and an acid whose composition is $C_2H_4(CO.OH)_2$, called *succinic acid*. Succinic acid reacts with bromine to form *dibromo-succinic acid*, $C_2H_2Br_2(CO.OH)_2$, and this acid is transformed into one of the four tartaric acids by boiling with water.

Inasmuch as ethylene can be formed from carbon and hydrogen, the series of reactions that begins with ethylene and ends with tartaric acid presents an example of the synthesis, in the laboratory, of a compound which is one of the characteristic products of living plants.

When the reactions whereby tartaric acid is synthesised from ethylene are considered in the light of other reactions that are similar to them, they lead to the structural formula $C_2H_2(OH)_2(CO.OH)_2$ for the tartaric acid that is thus obtained. The other reactions of this acid are altogether in keeping with this formula. Moreover all the purely chemical reactions of the acid are suggested by the following representation of the linkings of the atoms and atomic groups in the molecule :—

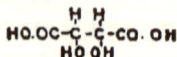

$$HO.OC-\underset{HO}{\overset{H}{C}}-\underset{OH}{\overset{H}{C}}-CO. OH$$

But the reactions of the other three tartaric acids also can be interpreted into the language of molecular structure, only by supposing that the arrangement of the atoms and atomic groups in the molecule of each of them is that which is presented by the above formula. We are then face to face with a new phenomenon. Four compounds exist having the same composition; the molecular weights of the four compounds are identical, and the four molecules are composed of the same numbers of the same atoms ; moreover, the reactions of the four compounds are so very similar that we are obliged to picture the arrangements of the atoms and atomic groups, that form the four molecules, by the same structural formula. Nethertheless the four compounds are not identical ; they differ in their crystalline forms ; two of them have the same solubilities in water, but the solubilities of the other two acids are different ; their melting points are not the same ; and they differ very widely in their optical activities. Although the chemical reactions of the four acids are very much alike, there are some differences. As regards optical activity : the acid prepared from succinic acid is inactive, but it can be resolved into equal weights of the dextro-rotatory and the laevo-rotatory acid ; the fourth variety of tartaric acid is optically inactive and cannot be resolved into the optically active acids. At this point we must inquire what is meant by *resolving the optically inactive acid* into the dextro-rotatory and laevo-rotatory varieties of tartaric acid. The *resolvable* inactive tartaric acid is known by the name *racemic acid.* If a solution of this acid in water is divided into two equal parts, if one half is neutralised by ammonia and the other half by soda, and if the two neutral liquids are mixed and the mixture is allowed to evaporate, crystals of ammonium-sodium racemate $(NaNH_4.C_4H_4O_6)$ are obtained. It is possible to separate these crystals, by hand-picking, into two sets, the difference between the crystals being the same as the difference between an object and its reflection in a mirror; a crystal of one kind is the reflected image of a crystal of the other kind; or, one may say, a crystal of one kind bears to a crystal of the other kind the same relation that a right-handed glove bears to a left-handed glove. By decomposing the crystals of one kind by the proper quantity of sulphuric acid, dextro-rotatory tartaric acid is obtained; and an equal quantity of laevo-rotatory tartaric acid is formed by decomposing the crystals of the other kind by sulphuric acid. The separation of inactive racemic acid into dextro-rotatory and laevo-rotatory tartaric acids can also be effected by other methods. There is no process known whereby the *non-resolvable* inactive tartaric acid (generally called *inactive tartaric acid*) can be separated, or resolved, into optically active acids.

There must be some differences in the structures of the molecules of the four tartaric acids; but the differences must be of a kind that is not expressed by the structural formulæ we have been using hitherto. In considering the structural formulæ we have employed as aids to forming clear conceptions regarding the connexions between the reactions of compounds and the arrangements of the atoms, and atomic groups, that form the molecules of these compounds, the careful reader will probably have observed that these formulæ make use of a certain convention which is palpably incorrect. The parts of a molecule must be arranged in three dimensions in space ; the molecule must have length, breadth, and thickness. But our structural formulæ have represented the parts of molecules as arranged only in two dimensions in space ; these formulæ present molecules as having length and breadth, but no thickness ; they make us think of the atoms, and atomic groups, as arranged all in one plane. May it be that the differences between the optical activities of the four tartaric acids are connected with different *spatial arrangements* of the same atoms, and groups of atoms, that form the molecules of these four compounds? In order to test this suggestion, it is necessary to adopt some convention, with the help of which we may form a clear working hypothesis regarding the connexions we are supposing to exist between the arrangements of the parts of molecules in three dimensions in space, and the properties of these molecules. At present I content myself with observing that a working hypothesis has been made, that formulæ have been constructed which enable us to connect the properties of many molecules with a spatial arrangement of their parts which can be thought about clearly, and that this hypothesis and these formulæ have led to many new conceptions of the greatest interest, and to new and fascinating suggestions, concerning the structure of those particles which are so

H

minute that the smallest piece of matter one can
see by the help of a first-rate microscope is about
one hundred million times larger than one of
them. I defer the consideration of this subject
until I come to speak of the sugars, in Chapter
IX.

CHAPTER VIII.

A FEW TECHNICAL APPLICATIONS OF
COMPOUNDS OF CARBON.

THE hydrocarbons methane (CH_4), ethane (C_2H_6),
and ethylene (C_2H_4) are constituents of coal-gas.
In the last chapter (p. 99), I drew attention to
the conditions under which coal-gas burns with a
luminous flame, and the conditions which destroy
the luminosity, while they increase the temperature, of the flame of ordinary gas. When coal-
gas is to be used for giving light, it must not be
allowed to mix freely with air before it is
ignited; when it is to be used as a source of heat,
the gas must be mixed with about fifteen times
its volume of air before it is burnt. But what is
the composition of coal-gas; and how is it manufactured? Let us briefly consider these questions.
If coal is strongly heated in an iron tube, closed at
one end, and fitted at the other end with a cork
through which passes a glass tube, gas will be
given off at the open end of the glass tube, and
this gas will take fire when a lighted taper is
brought near it. If the glass tube is bent downwards and passed through a cork which fits into

the mouth of a glass bottle, and another tube also passes through this cork into the air, tar and a watery liquid will collect in the bottle, and inflammable gas will issue from the open end of the exit-tube from the bottle. This experiment illustrates part of the process which takes place in a manufactory of coal-gas. Coal, usually a kind called *cannel coal* (because a piece of it burns like a candle), is placed in fire-clay retorts which are strongly heated in a furnace; the products are cooled, whereby tar, and water holding ammonia in solution, are condensed; the gas is washed (to remove ammonia), passed through purifiers (to remove carbonic acid gas, compounds of sulphur, etc.), and stored in gas-holders, from which it is sent, through the gas-mains, to the places where it is to be burnt. One ton of average cannel coal yields about 11,000 cubic feet of gas, about 13 gallons of gas-liquor (that is, tar and watery solution of ammonia), and about half a ton of coke which remains in the retorts. The number of compounds found in the products of the distillation of coal in a closed space is very great: these compounds comprise about 50 hydrocarbons; about a dozen compounds of carbon, hydrogen, and oxygen; about 15 compounds of carbon, hydrogen, and nitrogen; 9 or 10 carbon compounds that contain sulphur; and the three elementary substances, carbon, hydrogen, and nitrogen. After the products of distillation have been cooled, by passing through condensers, and the tar and ammoniacal liquor have thus been removed, the gas contains ammonia and various compounds of ammonia,

carbonic acid, sulphuretted hydrogen, carbon bisulphide, and small quantities of other compounds
of sulphur, besides the bodies which form the
mixture that is called coal-gas; most of the
ammonia and the compounds of ammonia are
removed by causing the gas to pass upwards
through a tower containing coke, or broken
flints, down which water trickles in thin streams;
the greater portion of the carbonic acid, the
whole of the sulphuretted hydrogen, and part of
the carbon bisulphide, are removed by passing
the gas through slaked lime, or through hydrated
oxide of iron and then through slaked lime.
The gas that now remains is a mixture of hydrogen, methane, ethylene, and other hydro-carbons,
carbon monoxide, a little carbon dioxide, nitrogen, oxygen, and small quantities of gaseous
sulphur compounds. The constituents of coal-gas
to which the luminosity of the flame of the gas is
chiefly due are the hydrocarbons other than
ethylene and methane. Part of the luminosity
is certainly caused by the glowing particles of
carbon produced by the decomposition of ethylene
(and to a smaller extent, of methane) in the
flame; but the other hydrocarbons, although
they are present only in small quantities, are
much richer in carbon than marsh gas and
ethylene, and they burn with a highly luminous
flame. No light is obtained by the burning of
the hydrogen and the carbon monoxide in coal-
gas; but the combustion of these gases produces
much heat, part of which is used in decomposing
the hydrocarbons with the formation of particles
of carbon. The nitrogen, oxygen, and carbonic

acid in coal-gas diminish the luminosity of the flame by dilution and cooling. The most objectionable impurities in coal-gas are carbon bisulphide and other sulphur compounds; when the gas is burnt, these compounds are oxidised to sulphurous and sulphuric acid, substances which are very injurious to health, and also to books and furniture. The coal-gas supplied in London must be perfectly free from sulphuretted hydrogen, and the quantity of sulphur in the other compounds of that element which are allowed to be present must not exceed $17\frac{1}{2}$ grains per 100 cubic feet of gas.

Among the products of the destructive distillation of coal are various hydrocarbons, belonging to the paraffin series, which are solids at ordinary temperatures. The composition of all the paraffins is given by the formula C_nH_{2n+2} (compare p. 103). The compound in which $n = 14$ ($C_{14}H_{30}$) melts at 4·5°C. [= 40°F.]; this compound, and all the paraffins higher in the series than this, are solids at the ordinary temperature. The solid, wax-like, substance that is known commercially as *paraffin* is a mixture of the higher hydrocarbons of the paraffin series. The liquid known as *paraffin oil* is a mixture of many hydrocarbons; the American oils consist chiefly of paraffins (C_nH_{2n+2}), while Russian oils generally contain also hydrocarbons of the composition C_nH_{2n}, and sometimes hydrocarbons belonging to the series C_nH_{2n-6}. The manufacture of paraffin-wax and paraffin-oil from shale is a process resembling that of the manufacture of coal-gas, in that both processes are based on the destructive distillation of carbonaceous minerals. When coaly shale is heated in a retort to low redness it is decomposed with the production of (i.) a gas which is not liquified at ordinary temperatures, (ii.) water holding a little ammonia in solution, (iii.) a thick oil which solidifies at the ordinary temperature, and (iv.) coke which remains in the retort. The gas is generally burnt, and the heat so produced is used in distilling fresh quantities of shale. The ammoniacal liquor is neutralised by acid, and the ammonia salt that is produced is used in other manufactures or as a manure. The oil is distilled, whereby it is separated into five portions:—(i.) naphtha, (ii.) burning oil, (iii.) light mineral oil, (iv.) lubricating oil, and (v.) paraffin-wax. *Naphtha* is the lightest and most volatile portion of the liquid products of the distillation of shale; it is generally separated, by re-distillation, into portions which have different specific gravities and boiling points, and are known by such names as *gasoline, benzine,* and *petroleum ether*. *Burning oil* is a mixture of hydrocarbons, from 30 to 80 per cent. of which belong to the olefine (or ethylene) series. *Mineral oil* and *lubricating oil* consist almost entirely of olefines; and *paraffin-wax* is a mixture of the higher (solid) members of the paraffin hydrocarbons. The district between Edinburgh and Glasgow is the seat of the manufacture of paraffin oils and solid paraffin from shaly coals. One ton of the shale which is distilled yields from 20 to 40 gallons of oil. In 1890, about 65 million gallons of crude oil were obtained; and this yielded about 2 million gallons of naphtha, about 19 million gallons of burning oil, about 10 million gallons of lubricating oil, and about 20,000 tons of solid paraffin.

In America and Russia, enormous quantities of oil are obtained by boring wells in the surface of the earth. These wells differ greatly in depth; in some places oil is found after boring to a depth of 40 or 50 feet, in other places the oil wells are nearly 3000 feet deep. Newly opened wells are generally "*gushers*"; that is to say, the pressure of the gases that are produced in the earth, along with the oil, is sufficient to cause the oil to flow, or in some cases to spout, from the openings of the wells. Some American newly opened "gushers" have poured forth as much as 8000 or 9000 barrels of oil per day; but the yield generally falls off rapidly, and in most cases it becomes necessary after a time to use pumps to raise the oil to the surface. One of the Russian wells spouted so violently that the outflow of oil was uncontrollable for three or four months; a column of oil from 100 to 300 feet high issued from the mouth of the well; the surrounding country was inundated; the workshops were nearly buried in the sand that was ejected with the oil; and from 100 to 200 million gallons of oil were lost. The Russian well ejected a column of oil and sand 400 feet high, and on windy-days the oil-spray was carried to a distance of eight miles.

The American and Russian crude oils are generally called *petroleum*. The processes whereby the petroleum is separated into naphtha, burning oil, lubricating oil, and paraffin-wax are essentially the same as those used for the separation of shale-oil into portions; they consist of distillation, washing with acid and alkali successively, and re-distillation. About 1000 million gallons of crude petroleum were produced in the United States in 1890, and about 76 million gallons were exported in that year. The quantity of crude petroleum exported from the Baku district in Russia amounted to about 3 million tons in 1890.

The oil-wells in America and Russia emit inflammable gases, besides those liquid compounds that form the petroleum of commerce. In the oil-bearing districts of Russia, especially in the neighbourhood of Baku, those inflammable gases have issued from the earth from time immemorial. Long ago, we do not know when or by whom, the issuing gas was ignited, and the *eternal fire* continued to burn for a great many centuries. An account of these fires was given, in 1754, by Mr Hanway, who was sent from England to arrange the conditions of a trade in oil from Baku to India *via* the Caspian Sea. Mr Hanway says :—

"What the Guebers, or Fire-Worshippers, call the Everlasting Fire is a phenomenon of a very extraordinary nature. The object of devotion lies about ten English miles north-east by east from the city of Baku, on a dry rocky land. There are several ancient temples built with stone, supposed to have been all dedicated to fire. Amongst others is a little temple at which the Indians now worship. Here are generally forty or fifty of these poor devotees, who come on a pilgrimage from their own country. A little way from the temple is a low cleft of a rock, in which there is a horizontal gap, two feet from the ground, nearly six long, and about three wide, out of which issues a constant flame, in

colour and gentleness not unlike a lamp that burns
with spirits, only more pure. When the wind blows, it
rises sometimes eight feet high, but much lower in still
weather. They do not perceive that the flame makes any
impression on the rock. This also the Indians worship,
and say it cannot be resisted, but if extinguished will
rise in another place. The earth round the place, for
about two miles, has this surprising property, that by
taking up two or three inches of the surface and applying
a live coal, the part which is so uncovered immediately
takes fire, almost before the coal touches the earth ; the
flame makes the soil hot, but does not consume it, nor
affect what is near it with any degree of heat. Any
quantity of this earth carried to another place does not
produce this effect. . . . If a cane or tube even of paper
be set about two inches in the ground, confined and closed
with earth below, and the top of it touched with a live
coal, and blown upon, immediately a flame issues without
hurting either the cane or paper, provided the edges be
covered with clay ; and this method they use for light in
their houses, which have only the earth for their floor ;
three or four of these lighted canes will boil water in a
pot, and thus they dress their victuals. . . . Lime is
burnt to great perfection by means of this phenomenon."

Gibbon tells us (in *The Decline and Fall*) that
in 624 A.D. Heraclius wintered 70 miles south of
Baku, and that he

"Signalised the zeal and revenge of a Christian em-
peror. At his command the soldiers extinguished the
fire and destroyed the temples of the Magi."

The process whereby acetic acid $(C_2H_4O_2)$ is
obtained from ethylic alcohol (C_2H_6O) has been
sketched in Chapter VI. (p. 83). The conver-
sion of ethylic alcohol, contained in certain fermented fruit-juices, into acetic acid is the main
chemical change that occurs in making *vinegar*.
Mere contact of alcohol, or an alcoholic liquor,
with oxygen does not suffice to effect the oxida-

tion of the alcohol to acetic acid ; but if a small quantity of the minute fungus called *mycoderma aceti* is present, the oxidation proceeds. This little plant grows in a liquid which contains albuminous bodies and certain mineral salts, provided there is free access of air ; and if the liquid also contains 10 per cent., or less than 10 per cent., of alcohol, the plant absorbs the alcohol slowly, and causes its oxidation to acetic acid by the oxygen which is taken from the air by the growing fungus. As the germs of this acetifying fungus are always present in the air, wine or beer soon begins to turn sour when it is kept in an open vessel. The most favourable conditions for the acetification of an alcoholic liquor are these : the presence in the liquid of plenty of food for the fungus ; the presence of not more than 10 per cent., nor less than 4 per cent., of alcohol ; the exposure of a large surface of the liquid to the free access of air ; and the maintenance of the temperature at about 25° C. [= 77° F.]. There are two processes whereby vinegar is manufactured. In the older process, the alcoholic liquid—generally poor wine about a year old, or a fermented infusion of malt—is kept for many days in large casks, made of beech-wood, which have been soaked in vinegar. In the quick vinegar process, an alcoholic liquid is caused to trickle, in very thin streams, over beechwood shavings, or purified charcoal, which have been " soured " by immersion in hot vinegar, while a current of air passes through the shavings, or the charcoal, in the direction opposite to that taken by the spray of alcoholic liquid.

The casks employed in the slow process contain from 50 to 100 gallons of liquid apiece ; each is pierced by two holes, one to admit air, and the other for pouring in, or withdrawing, liquid. The casks are filled to one-third with vinegar, and the temperature is kept at about 25° C. [= 77° F.] ; after eight days about ten pints of wine are added, and after another eight days about ten pints more, and so on, until the casks are two-thirds full. When fourteen days have passed, a portion of the liquid is drawn off, and more wine is poured in. Each cask produces annually a quantity of vinegar equal to about twice its own capacity. The casks are thoroughly cleaned about once in six years. It is said that a good cask will last for five and twenty years. The manufacture of malt vinegar is very much like that of wine vinegar. Sometimes the casks are placed in rows in the open air ; in such a case the operation generally begins in the spring, and is finished in about three months.

The liquors that are acetified by the quick process are generally prepared by mixing poor brandies with water and some vinegar, and adding bran or rye to give food to the vinegar fungus. Sometimes diluted brandy, or whiskey, is mixed with fermented infusion of malt ; in other cases an infusion of barley-meal and wheat-meal is fermented ; molasses or honey is added occasionally to give a rich colour to the vinegar. The acetification is conducted in large vats, which are furnished with perforated false bottoms placed about 18 inches above the true bottoms, and fitted near the tops with wooden discs which are pierced by a great many very small holes; there are also a few holes sloping downwards in the sides of each vat, below the false bottom. The space between the false bottom and the upper disc is nearly filled with beech-wood shavings, or with pieces of charcoal which have been freed from saline impurities by soaking in acid and washing. Through each hole in the upper disc is suspended a thread of twisted cotton yarn, the lower end of which touches the shavings, or the charcoal. The shavings (or charcoal) are soaked in hot vinegar for a day or two. The liquor is poured on to the upper disc, and trickles slowly down the twisted threads and then through the shavings, or the charcoal; when the shavings have become coated with the vinegar fungus, the oxidation of the alcohol in the liquor proceeds fairly rapidly, and the temperature rises in the interior of the vat to about 37° C. [= 98° F.]: one effect of this heating is to cause a current of air to enter, by the holes in the sides of the vat beneath the false bottom, and, in passing upwards, to come in contact with the descending thin stream of liquid, and thus to aid the oxidation of the alcohol in the liquid. The vinegar is drawn off by a tap placed at about an inch above the bottom of the vat. If the alcoholic liquor contains about 4 per cent. of alcohol, the acetification is complete when the liquid reaches the bottom of the vat; liquids richer in alcohol must be passed through the apparatus two or three times.

Besides acetic acid, which is the main product of the chemical changes that occur in making vinegar, there are formed small quantities of ethylic acetate, and some other ethereal salts; it is to the presence of these compounds that the odour of vinegar is chiefly due. The quantity of acetic acid should never be less than 5 per cent. in genuine vinegar, but wine vinegar sometimes contains as much as 12 per cent. Besides acetic acid, vinegar contains small quantities of alcohol, sugar, and various substances extracted from malt or wine-juice, in addition to chlorides, acetates, sulphates, and phosphates, of various metals, the principal of which are sodium, potassium, and calcium. It is legal to add one part of sulphuric acid to 1000 parts of vinegar.

An *artificial vinegar* is prepared by mixing acetic acid with water, and adding a little burnt sugar, and a trace of ethylic acetate to give an agreeable odour.

The only other technical application of compounds of carbon which I shall notice in this chapter is that wherein the ethereal salts, glyceryl stearate, palmitate, and oleate, are decomposed, by alkalis, to produce soaps and glycerin. We have already considered the chemical changes that occur when caustic potash, or soda, is boiled with an ethereal salt; the products are a potassium, or sodium, salt of the acid the radicle whereof formed part of the ethereal salt used, and an alcohol, that is a hydroxide of the ethereal radicle of the salt that has been decomposed. (For a detailed account of these changes see the last chapter, pp. 94, 95.) The reactions that occur when glyceryl stearate, palmitate, or oleate, is decomposed by steam are

very much like those which take place when the
decomposition is effected by caustic potash; only,
in place of obtaining potassium stearate, palmitate, or oleate, we obtain stearic, palmitic, or
oleïc acid.   When glyceryl stearate is used, the
reaction may be expressed thus:—

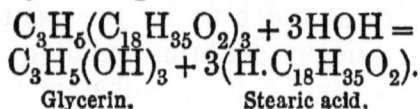

$$C_3H_5(C_{18}H_{35}O_2)_3 + 3HOH =$$
$$C_3H_5(OH)_3 + 3(H.C_{18}H_{35}O_2).$$

Glycerin.   Stearic acid.

For the sake of comparison, the reaction between
glyceryl stearate and a boiling solution of potash
is repeated here :—

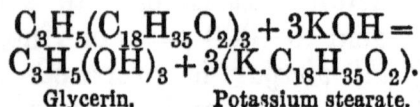

$$C_3H_5(C_{18}H_{35}O_2)_3 + 3KOH =$$
$$C_3H_5(OH)_3 + 3(K.C_{18}H_{35}O_2).$$

Glycerin.   Potassium stearate.

The decomposition of the glyceryl salts of
palmitic, stearic, oleïc, and other acids, for the
purpose of making soaps, is conducted in large
metal vats, holding from 10 to 40 tons of
material; the fatty matter is melted, weak soda
ley (solution of caustic soda) is run in, and the
whole is heated to boiling by steam; after a time
more concentrated soda ley is added, and boiling
is continued until the fat has been saponified;
common salt, or a concentrated solution of common salt, is then added, because soap is insoluble in concentrated brine ; as the contents of
the vat cool, the soap separates to the top, as
a curd, and the glycerin dissolves in the lower
watery layer.   The curd is separated, and boiled
with successive quantities of soda ley until it has
a distinctly alkaline taste ; the curd is again
allowed to separate, and is then run into cooling

frames made of iron or wood. The semi-fluid material is often mixed with scents, antiseptic substances, or such salts as silicate of soda or sulphate of soda; the presence of these salts gives a finer texture to the soap, and also enables it to take up a large quantity of water and yet remain solid. Sometimes the soap is '*fitted.*' In this process the soap is heated by means of wet steam, and then allowed to rest for some days, when a separation occurs into three layers: the upper layer is a frothy soap known as '*fob*'; the lowest layer contains various impurities, and water, mixed perhaps with more or less caustic soda; and the middle layer consists of the '*neat*' soap, which is run into cooling frames. A fitted soap may contain 30 or 40 per cent. of water, but it is almost quite free from caustic soda; curd soap, on the other hand, often contains considerable quantities of alkali. If such fats as linseed, poppy, or hempseed, oil are boiled with just sufficient potash ley to complete the process of saponification, and the whole mass is allowed to cool, a soft, homogeneous jelly is produced which contains the whole of the glycerin formed in the saponification; such a soap is very soft and dissolves easily in water; it generally, however, contains free alkali.

Some of the finest toilet soaps are obtained by drying carefully made soap, then dissolving it in alcohol, and distilling off the spirit; the residue is more or less transparent; as the soap sets it is mixed with a little glycerin.

Soap is fairly soluble in water, but it is insoluble in water containing a certain, not very large, quantity of saline matter. As soaps made from cocoanut oil, and palm-kernel oil, are more soluble in a watery solution of saline matter than other soaps, the products of saponifying these oils by potash are sold as *marine soaps*, because they give a lather with salt water. If soap made in this way is mixed with such salts as silicate or sulphate of soda, it is possible to add a very large quantity of water and yet to obtain a firm mass ; marine soaps sometimes contain 80 to 85 per cent. of water and salts, and only 15 to 20 per cent. of genuine soap.

A theoretically perfect soap consists of potassium, or sodium, salts of certain fatty acids, mixed together. Contact with water breaks up the soap into salts which contain more of the acidic radicle relatively to the quantity of potassium or sodium present, and some potash, or soda ; and the detergent value of soap is largely due to the small quantity of alkali thus produced. If the water which comes into contact with soap contains chalk, or gypsum, or carbonate or sulphate of magnesium, then a reaction occurs between the soap and the salt in the water, whereby a calcium (or magnesium) salt of the acidic radicle of the soap, and a potassium (or sodium) salt of the acidic radicle of the salt that is present in the water, are produced. Supposing the soap to consist of potassium stearate only, and the water to contain calcium carbonate (chalk), then the reaction may be thus expressed in an equation—

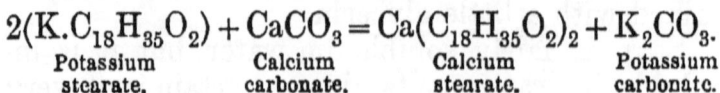

$$2(K.C_{18}H_{35}O_2) + CaCO_3 = Ca(C_{18}H_{35}O_2)_2 + K_2CO_3.$$

| Potassium stearate. | Calcium carbonate. | Calcium stearate. | Potassium carbonate. |

When the reaction between the soap and the salts in the water is finished, but not until then, the ordinary action between a soap and pure water begins, and a lather is produced. All *"hard"* waters contain chalk, or gypsum, or sulphate or carbonate of magnesium, in solution; hence, in washing with a hard water, a comparatively large quantity of soap must be used before a lather is produced.

CHAPTER IX.

SUGARS, STARCHES, AND CELLULOSE.

THE juices of the sugar-cane, beetroot, certain palms, the maple, and the *sorghum*, contain a compound of carbon, hydrogen, and oxygen, which has the composition $C_{12}H_{22}O_{11}$. All plants contain starch, $C_6H_{10}O_5$. And the main constituent of all vegetable tissues is cellulose, $C_6H_{10}O_5$. A sugar, whose composition is expressed by the formula $C_6H_{10}O_6$, is found in the juices of fruits, and in honey.

Cane-sugar, or *saccharose* $(C_{12}H_{22}O_{11})$ is prepared by crushing sugar-cane, or beetroots, neutralising the juice by lime, adding sulphurous acid (to prevent fermentation), evaporating, and crystallising. The raw sugar is refined by dissolving it in water, filtering, removing colouring matter by means of animal charcoal, evaporating in vacuum-pans, and separating the solid sugar from the syrup by centrifugal machines. *Starch*

I

is obtained by steeping potatoes in water, washing, rasping, and straining, and allowing the
starch to settle; the starch is then washed,
drained, and dried. Rice-starch is manufactured
by macerating rice with a dilute solution of soda
or potash, whereby gluten is removed, draining,
washing, grinding, and sifting; in some processes
the gluten is removed by a process of fermentation, followed by washing and treatment with
much diluted acid. The variety of fruit-sugar
known as *glucose* ($C_6H_{12}O_6$) is generally manufactured from the starch of sago, maize, or rice,
by heating with dilute sulphuric acid, neutralising by chalk, separating from calcium sulphate,
decolourising by animal charcoal, and evaporating to a syrup. *Cellulose* is made from vegetable
tissues by treatment with alkali and some weak
oxidiser; for instance, by treating cotton with
bleaching powder.

These four compounds, cane-sugar, fruit-sugar,
starch, and cellulose, belong to the class of compounds called *carbohydrates*. The formula of each
contains six atoms, or a whole multiple of six
atoms, of carbon, and always twice as many
atoms of hydrogen as of oxygen; in other words,
the compositions of the compounds are expressed
by the formula $nC_6.mH_2O$. Although the weights
of hydrogen and oxygen in these compounds are
in the same ratio as in water, the substances are
not compounds of carbon with water (as the name
carbohydrate implies); their reactions negative
this view of their constitution.

Cane-sugar is the most important representative
of the group of carbohydrates called *saccharoses*,

all of which have the composition $C_{12}H_{22}O_{11}$. Fruit-sugar belongs to the group of *glucoses*, $C_6H_{12}O_6$. And starch and cellulose are *amyloses*, $C_6H_{10}O_5$. These formulæ are the simplest expressions that can be given of the compositions and some of the reactions of the four compounds, but they are not necessarily molecular formulæ; the formulæ which tell the numbers of atoms of carbon, hydrogen, and oxygen, in the molecules of the compounds may be multiples of these simplest expressions. There is, however, good reason to regard the formulæ $C_{12}H_{22}O_{11}$ and $C_6H_{12}O_6$, given to cane-sugar and fruit-sugar, respectively, as molecular formulæ.

The examination of the reactions of the glucoses has been carried much further than that of the reactions of the saccharoses or of the amyloses. There are at least eight sugars to all of which the formula $C_6H_{12}O_6$ must be given ; moreover the reactions of these eight sugars shew that the structural formula $CH_2OH.(CHOH)_4.HCO$ must be assigned to each of them. Now it is not possible to form eight modifications of this expression if the assumption is made that the atoms are arranged in only two dimensions in space. It has been noticed already (see pp. 112, 113) that the hypothesis that molecules have only length and breadth and no thickness served admirably to group together many facts concerning the reactions of carbon compounds, but that the hypothesis broke down in some cases, for instance in the case of the four tartaric acids (see pp. 111, 112). We have now another set of facts which refuse to be brought into order by the use of the hypothesis that worked so well for a time. We are forced to attempt to form expressions which shall represent the atoms of carbon, hydrogen and oxygen as arranged in three dimensions in space in those groups which are the molecules of the various glucoses.

It is of course impossible to form realistic presentments of the tri-dimensional arrangements of the atoms in molecules; all we can do is to endeavour to construct an hypothesis which shall be the framework wherein our knowledge of the observed reactions of the compounds under consideration may be set, shall bind that knowledge into a consistent whole, and shall indicate the directions wherein new attacks, likely to prove successful, may be made on the problem of the connexions between the compositions and the reactions of molecules.

The glucoses resemble some of the tartaric acids in one respect; they are optically active compounds when dissolved in water: solutions of some of them rotate the plane of polarisation of a ray of light to the right hand, and solutions of some are lævo-rotatory. An aqueous solution of cane sugar is dextro-rotatory; ordinary starch which has been dried in the air is insoluble in water, but there is a variety of starch called *soluble starch*, and an aqueous solution of that compound rotates the plane of polarisation of a ray of light in the direction opposite to that in which the hands of a watch move; cellulose is insoluble in water.

The hypothesis that is used in framing tri-dimensional formulæ for such compounds as the

glucoses and the tartaric acids rests on the conception of *the asymmetric carbon atom.*
Let an atom of carbon be in direct union with
four other atoms, two of which are identical, and
the two others are also identical but different
from the first pair; the composition of such a
molecule will be expressed by the symbol
$CR_1R_1R_2R_2$. The compound CH_2Cl_2 (dichloro-
methane) is an example of this arrangement.
Now it is possible to assign two different formulæ to this molecule, using the ordinary
hypothesis that the atoms are arranged in
two dimensions in space; these formulæ are

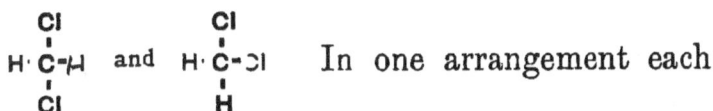

Cl   Cl
H·C·H   and   H·C·Cl   In one arrangement each
Cl   H

hydrogen atom has a chlorine atom on either
side of it; in the other arrangement each
hydrogen atom has for its immediate neigh-
bours an atom of chlorine and an atom of hydrogen. But observed facts tell that molecules
of this composition never exist in more than
one modification: there is only one dichloro-
methane, not two dichloromethanes as there
ought to be if the ordinary view of the arrangement of atoms were sufficient. We want then to
picture the arrangement in space of a molecule
composed of an atom of carbon united to four
other atoms, or atomic groups, in such a way as
shall bring our theoretical conception of this
arrangement into keeping with the observed
facts. Now suppose the carbon atom to be
placed in the centre of a regular tetrahedron,

and each of the four other atoms to be placed
at one of the summits of the tetrahedron ; we
have the arrangement pictured thus :—

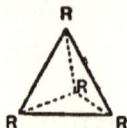

where each R stands for an atom, or atomic
group, in direct union with the atom of carbon
supposed to be in the centre of the tetrahedron.
Suppose the four atoms, or groups, in connexion
with the carbon atom to be different (suppose
them, for instance, to be an atom of hydrogen,
an atom of chlorine, an atom of bromine, and
an atom of iodine) ; then the arrangement would
be pictured by one of the two following figures,
and the image of this arrangement in a mirror
would be represented by the other figure :—

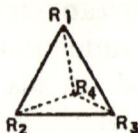

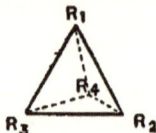

We have here two arrangements like a right-
handed and a left-handed glove. Just as it is
impossible to lay a pair of gloves together, both
palms upward, or both backs upward, without
bringing the thumbs on different sides, so it is
impossible to lay one of the arrangements shewn
in the figures on the other so that R_1 shall be
superposed on R_1, R_2 on R_2, R_3 on R_3, and R_4
on R_4. The properties of one of the molecules
pictured by, say, the first of the two figures will
differ from the properties of the molecule pictured
by the other figure.

When two compounds have the same composition but different properties, and the two molecules are composed of the same numbers of the same atoms, one compound is said to be an *isomeride* of the other ; and the existence of two (or more than two) such compounds is said to be a case of *isomerism* (from two Greek words signifying *equal parts*). In the case before us, the compound symbolised by one of the tetrahedral figures is called a *geometrical isomeride*, or sometimes a *mirror-isomeride*, of the other.

Now consider the case of an atom of carbon in direct union with four other atoms, or groups of atoms, two of which are the same. Representing these atoms, or groups, by $R_1 R_1 R_2$ and R_3, we can picture the tetrahedral arrangement of all the atoms by one of the following figures, and the mirror-image of that arrangement by the other figure :—

One of these figures can be superposed on the other : take the second figure, turn it round so that the R_1 at the top of the figure is brought where the R_2 was before the figure was turned round ; then the result is identical with the first figure. In such a case as this, that is in a molecule composed of an atom of carbon united with four other atoms (or groups) two of which are the same, isomerism cannot occur if the hypothesis we are working on is a satisfactory method of symbolising facts. There is no case known of

the existence of more than one modification of a
compound whose molecular composition is expressed by the symbol $CR_1R_1R_2R_3$. There are
cases known of isomerism shewn by compounds
whose composition is expressed by the symbol
$CR_1R_2R_3R_4$.

Turn back for a moment to the two figures
placed side by side on p. 134. The fact that
neither figure is superposable on the other is
sometimes expressed by saying there is no *plane
of symmetry* in either figure. It is this notion of
symmetry which underlies the expression, I am
now trying to explain, *the asymmetric carbon atom.*
If a single atom of carbon is in direct union with
four other atoms, or atomic groups, all of which
are different, then a compound is produced which
can exist in two modifications; or, it would be
more correct to say, two compounds are possible
both having the same composition. But if a
single atom of carbon is in direct union with four
other atoms (or groups), two, or three, of which
are the same, then the compound that is formed
does not exist in more than one modification.
The only way we have been able to think clearly
about these facts (the reader should notice that
it is almost impossible to state the bare facts
except in terms of a theory of the structure of
matter) is by picturing to ourselves the arrangement of the atoms of the two isomeric molecules,
$CR_1R_2R_3R_4$, as like a regular tetrahedron with
the atom of carbon in the centre, and each atom,
or atomic group, at one of the summits. And as
this is an unsymmetrical arrangement, inasmuch
as the mirror-image of this arrangement cannot

be superposed on the original configuration, the atom of carbon, which is thought of as the central pivot whereon the other atoms are hung, is spoken of as an *asymmetric atom.* In some cases then we think of the existence of two, or more than two, compounds of carbon with the same composition and the same molecular weight as dependent on the presence of asymmetric carbon atoms in the molecules of these compounds ; and by the expression, *an asymmetric carbon atom*, we mean an atom of carbon in direct union with four atoms (or atomic groups) no one of which is the same as any other. The only way we have at present of connecting the exhibition of isomerism with the presence of asymmetric carbon atoms in molecules is by likening the configurations of these molecules to tetrahedral figures, with an asymmetric carbon atom in the centre of each, and the four atoms, or groups, that are in direct union with this asymmetric atom at the four summits of each figure. Of course we are sure that the molecules are not really tetrahedral arrangements with an atom of carbon in the centre, for we know that a molecule must be an exceedingly complicated structure, and that the parts of every molecule must be performing regulated movements. Nevertheless we must frame some conception of the arrangement of the parts of molecules, and in order to form a conception which shall be useful in furthering exact knowledge of the connexions between composition and properties, we must employ a tool the trick of which we have learned and can use. The only tool that has been found suitable for the work to be done is that whose mechanism I have been trying to describe.

The reactions of every compound which is optically active in solution (that is to say, which rotates the plane of polarisation of a ray of light) indicate the existence of at least one atom of carbon in the molecule in direct union with four different atoms, or groups of atoms. The exhibition of optical activity by a compound in solution seems then to be associated with the presence of asymmetric carbon atoms in the molecule of the compound. We found (Chapter VII., p. 109) that four tartaric acids exist, that the reactions of all are expressed by the formula $C_2H_2(OH)_2(CO_2H)_2$, that one of these acids is dextrorotatory in solution, another is lævorotatory, another is optically inactive but can be resolved into equal weights of the dextrorotatory and the lævorotatory acids, and that the fourth is optically inactive in solution and cannot be resolved into the active modifications. Writing the formula $C_2H_2(OH)_2(CO_2H)_2$ in full, we have the following expression :—

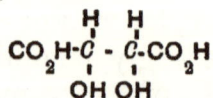

$$CO_2H\text{-}\overset{\overset{\displaystyle H}{|}}{C}\text{ - }\overset{\overset{\displaystyle H}{|}}{\underset{\underset{\displaystyle OH}{|}}{C}}\text{-}CO_2H$$
$$\underset{OH}{|}$$

The molecule of tartaric acid contains two atoms of carbon each of which is in direct union with four different atoms or atomic groups, namely, with H, (OH), (CO_2H), and $(C.H.OH.CO_2H)$; that is, the molecule contains two asymmetric carbon atoms (the symbols of these atoms are printed in italics in the formula). Now if this formula must be assigned to each of the four tartaric acids, it is evident that a compound may be optically inactive although its molecule contains two asymmetric carbon atoms, and that an optically active compound may have the same composition as another which is inactive and, like the active compound, contains a pair of asymmetric carbon atoms.

To attempt to follow in detail the working out of the hypothesis of geometrical isomerism as it is applied to the cases of the four tartaric acids would be out of place in this book; nevertheless a slight outline of the application of the hypothesis may be given. The conception that is formed of the structure of the molecule of tartaric acid is that of two tetradedra with one summit of one joined to one of the other; one asymmetrical carbon atom is thought of as placed at the centre of each tetrahedron, and the atomic groups are supposed to be placed at the remaining six summits. Two of the possible arrangements of those groups are shown in the following figures :—

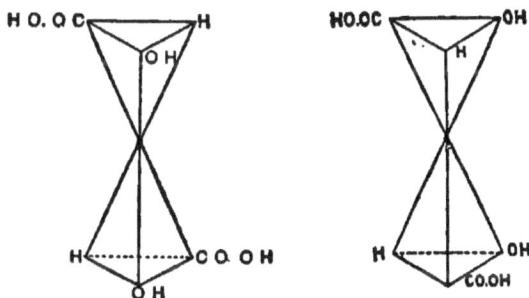

One of these figures bears to the other the same relation as an object bears to its image in a mirror. If one of these molecules is dextro-rotatory the other will be lævorotatory. If we suppose a dextrorotatory molecule joined to a lævorotatory molecule the right-handed optical activity of one will be neutralised by the left-handed activity of the other; the compound molecule will be optically inactive, but resolvable into a dextrorotatory and a lævorotatory mole-cule. The existence of two optically active, and one inactive but resolvable, modification of tar-taric acid is thus in keeping with the hypothesis. It is also possible to represent the molecule of tartaric acid by such a modification of the figures given above that one-half of the molecule shall be the mirror-image of the other half. One half of such an arrangement would be dextrorotatory and the other half lævorotatory; the dextroro-tatory part would neutralise the other, and the whole molecule would be optically inactive; but this modification of tartaric acid would not be resolvable into two optically active acids, because splitting a molecule into parts is the same thing as changing the substance into other substances altogether different from it.

The hypothesis which rests on the conception of the asymmetric carbon atom has been applied to the glucoses, and has been found sufficient to bring the reactions of this class of sugars under one general principle, to elucidate the relations which experiments have shown to exist between these compounds, and to indicate many reactions which have afterwards found experimental veri-fication.

We have learned in this chapter that the influ-ence of carbon atoms on the properties of the

molecules whereof they form parts are profoundly
modified by the relations which exist between
these atoms and the other atoms that, with them,
constitute the molecules.   In all its wanderings
carbon preserves a certain sameness of character,
but it is enormously influenced by its companions.
A very small change in the spatial arrangement
of a few atoms of which carbon is one is sometimes accompanied by a great change in the
properties of the whole group of atoms.   The
facts concerning the properties of such compounds
as the tartaric acids and the glucoses drive home
the importance of the study of the finer relations
between the parts of compounds.   No amount of
examination of the elements that form tartaric
acid, or glucose, could in the least prepare us to
expect that any collocation of these elements
should possess the properties which characterise
the compounds that are formed by their union.
The composition of every compound is absolutely
unalterable ; the properties of every element,
taken by itself, are fixed and unchangeable ; yet
join together the same elements, and the same
quantities of the same elements, and you produce compounds in many cases very unlike one
another.   As investigation advances, the problem
of the connexions between composition and properties becomes finer and more elusive, until we
come to bodies which are identical so far as we
can test them in the laboratory but produce completely different effects on living organisms.

CHAPTER X.

Pure.

In Chapter III. it was said that most of the compounds of carbon belong to one or other of two main classes; those which bear a general resemblance to the paraffins and are derived from these hydrocarbons, and those which are derived from the hydrocarbon benzene and are related to that compound in their reactions. Hitherto we have been concerned with *paraffinoid compounds*; in this chapter we shall have to deal with a few *benzenoid compounds*. The common fats are included in the first class of compounds, and all the members of that class are often called *fatty compounds*; inasmuch as those substances which give aromatic odours to certain plants belong to the benzenoid class, the name *aromatic compounds* is commonly applied to all the members of that class.

• A vast number of compounds is derived from the hydrocarbon benzene; in this chapter we shall consider a few of these. Benzene is composed of carbon and hydrogen united in the proportion of 92·3 per cent. of carbon to 7·7 per cent. of hydrogen : as the atomic weight of carbon is 12, and that of hydrogen is 1, it follows that there is the same number of atoms of carbon as of hydrogen in the molecule of benzene [$\frac{92·3}{12} = 7·7$; $\frac{7·7}{1} = 7·7$]. The molecular weight of this compound is found to be 78; hence the molecule is composed of 6 atoms of carbon united to 6 atoms of hydrogen $[(6 \times 12) + (6 \times 1) = 78]$, and the molecular formula of the compound is C_6H_6. No aromatic compound is known containing less than 6 atoms of carbon in its molecule.

Benzene can be obtained, by a somewhat indirect process, from gum benzoïn; as benzene is an oily substance, the name *benzole* (probably abbreviated from *benzoïn oleum*) was originally given to the hydrocarbon, and this is the name by which the compound is generally known in commerce both in this country and abroad. Benzene is manufactured from coal-tar, by repeatedly distilling the lower boiling portions until a liquid boiling at about 80° C. [= 176° F.] is obtained. The hydrocarbon is a clear, colourless, limpid liquid, somewhat lighter than water bulk for bulk; it freezes to colourless crystals at a temperature a little higher than that of the freezing point of water.

When benzene is mixed with concentrated nitric acid, an oil smelling like oil of bitter almonds is produced; this oil has the composition $C_6H_5.NO_2$, and is called *nitrobenzene*. By subjecting this compound to the action of iron filings and acetic acid it is converted into *aniline*, $C_6H_5NH_2$, from which is obtained a series of compounds called the aniline colours. Another compound called *phenol*, or often *carbolic acid*, is obtained from benzene by a series of reactions; this compound has the composition $C_6H_5.OH$. One of the products of the oxidation of benzene is *benzoic acid* $C_6H_5.CO_2H$; and from this acid is obtained another named *salicylic acid*,

$C_6H_4.OH.CO_2H$. I propose to consider these compounds briefly; always with the object of tracing connexions between the compositions and the properties of definite kinds of matter.

There are three statements concerning the reactions of benzene and its derivatives which must be insisted on. (1) In almost all of its reactions benzene acts as a saturated compound; that is to say, it shews very little liking for the addition of other elements to itself, it is generally ready to exchange hydrogen for other elements. For instance, it is easy to form C_6H_5Cl or $C_6H_4Cl_2$, $C_6H_5.NO_2$ or $C_6H_4(NO_2)_2$, from C_6H_6; but the formation of $C_6H_6Cl_6$ proceeds only very slowly and under the influence of sunshine, and when this compound is produced it fairly readily changes to $C_6H_3Cl_3$ and hydrochloric acid (HCl). (2) The general result of the action of reagents on the derivatives of benzene is to produce some compound, or compounds, containing six atoms of carbon in the molecule; such compounds break down into simpler bodies only by the prolonged action of energetic reagents. (3) The compound C_6H_5Cl is called *monochloro-benzene*, and the compound $C_6H_4Cl_2$ *dichloro-benzene*; there is only one monochloro-benzene, but there are three different dichloro-benzenes. There are also three dibromo-benzenes ($C_6H_4Br_2$), three dinitro-benzenes [$C_6H_4(NO_2)_2$], and generally three isomeric compounds of the composition $C_6H_4R_2$ where R is an atom, or an atomic group, capable of replacing one atom of hydrogen in a molecule.

How are these facts to be suggested in a

structural formula for benzene ? How are we to represent the arrangement of six carbon atoms and six hydrogen atoms so that the facts concerning the reactions of the molecule C_6H_6 may be implicitly contained in the formula which expresses the supposed arrangement ? Thirty-four years ago (in March, 1865) the German naturalist Kekulé was living in London. He had been thinking eagerly for some time about the structure of the molecules of the benzene compounds. As he was riding, one day, on the top of a Clapham omnibus he saw in his mind's eye the arrangement of the atoms in the molecule C_6H_6; the problem was solved. Kekulé pictured to himself six atoms of carbon arranged so that each was in direct union with two other atoms of carbon and with one atom of hydrogen. The simplest way of presenting this arrangement on a plane surface, using the ordinary conventions, is perhaps the following :

$$\begin{array}{c} H \\ H\text{-}C\text{-}C\text{-}C\text{-}H \\ H\text{-}C\text{-}C\text{-}C\text{-}H \\ H \end{array}$$

Kekulé preferred to think of the six carbon atoms as forming an hexagonal figure ; thus

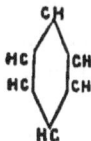

$$\begin{array}{c} CH \\ HC \quad CH \\ HC \quad CH \\ HC \end{array}$$

And since that time chemists have been accustomed to speak of *the hexagon-formula of benzene.* The six atoms of carbon are thought of as forming a compact, stable *nucleus*, and the six atoms of hydrogen as attached to this nucleus

K

each to one of the six carbon atoms. The
derivatives of benzene are represented as formed
by removing one, or more, atoms of hydrogen
from the molecule C_6H_6, and putting other atoms,
and groups of atoms, in their places, the six
carbon nucleus remaining intact. The general
action of reagents on these derivatives of benzene
is pictured as resulting in the breaking off of
some, or all, of the *side chains*, without (in most
cases) decomposing the nucleus of six carbon
atoms. If six atoms of chlorine are added to the
molecule C_6H_6, it should not be possible to add
any more chlorine; because the molecule of
benzene hexachloride would have this structure:

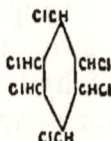

and experience tells that an atom of carbon
cannot hold directly to itself more than four
other atoms of any kind in a molecule. Finally,
consider the structure of the isomeric dichloro-
benzenes which ought to exist if Kekulé's hypothesis is a sufficient translation of the facts into
the language of molecular structure. There may
be three, but not more than three, dichloro-benzenes; and the formulæ of these three isomeric
compounds are:

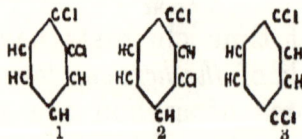

In the first of these structures the chlorine

atoms are attached to contiguous carbon atoms ; in the second, one atom of carbon comes between the two carbon atoms which hold chlorine atoms bound to themselves ; and in the third structure, two carbon atoms intervene between the carbons which are attached to atoms of chlorine. Any other arrangement of $C_6H_4Cl_2$ in terms of Kekulé's hypothesis, would be identical with one or other of these three. For, suppose the carbon atoms are numbered from the top of the hexagon in the direction wherein the hands of a watch move ; we have this arrangement :

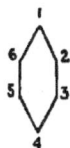

Then, calling the first $C_6H_4Cl_2$ 1:2 dichloro-benzene, the second 1:3, and the third 1:4 dichloro-benzene, inspection of the formulæ will shew that if the chlorine atoms were attached to the carbon atoms 1 and 6, the compound so produced would be the same as that formed by attaching the chlorine atom to the carbon atoms 1 and 2, and that the position 1:5 is the same as 1:3.

Kekulé's formula for benzene then summarises, in a special language, the chief reactions of this compound and its derivatives ; and investigation has shown that the formula suggested by Kekulé has been most fruitful in suggesting lines of advance by pursuing which much accurate knowledge has been gained not only concerning benzene compounds, but also concerning the main problem of Chemistry, which, as has been said so often, is to elucidate the connexions between composition and properties.

It should be noted that the hexagon-formula does not pay any heed to the arrangement of the atoms in three dimensions in space. The formula, which is a sentence in a language, has proved sufficiently elastic to include many facts discovered since it was constructed; and when it was found necessary to adopt an expression for the structure of certain benzene compounds, which includes in itself the conception of the tri-dimensional arrangement of atoms, Kekulé's formula was not abandoned, but only modified. Even a slight study of the meaning and purpose of the benzene hexagon formula will enable the reader to realise to some extent what can be done by the use of hypotheses in natural science, and how absolutely necessary it is to frame and constantly employ hypotheses if advance is to be made in the accurate investigation of natural occurrences. Incidentally, it may be added that Kekulé's ride on the Clapham 'bus has been worth millions of pounds to chemical manufacturers.

The relations between benzene, phenol, nitrobenzene, aniline, benzoic acid, and salicylic acid are suggested by the following structural formulæ :—

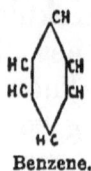

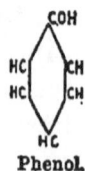

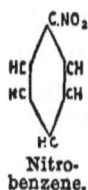

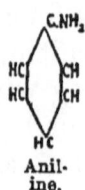

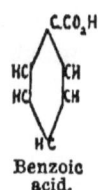

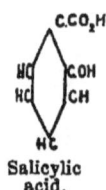

Benzene.	Phenol.	Nitro-benzene.	Aniline.	Benzoic acid.	Salicylic acid.
CH	$C.OH$	$C.NO_2$	$C.NH_2$	$C.CO_2H$	$C.CO_2H$
HC CH HC CH HC	HC CH HC CH HC	HC CH HC CH HC	HC CH HC CH HC	HC CH HC CH HC	HC $C.OH$ HC CH HC

Let us endeavour to read something of the meanings of one or two of these formulæ.

So far as composition is concerned, the relation of phenol to benzene is similar to that of methylic alcohol to methane, and to that of ethylic alcohol to ethane. Methane and methylic alcohol were briefly examined in Chap. V. (pp. 65-69), and ethane and ethylic alcohol in Chap. VI. (pp. 83-85). The substitution of the atomic group OH for one of the hydrogen atoms in the molecule of either of the hydrocarbons CH_4 or C_2H_6 is accompanied by a change from the properties of a hydrocarbon to those of an alcohol. Benzene is a hydrocarbon, and the change of composition that occurs when benzene becomes phenol consists in the substitution of the atomic group OH for one of the hydrogen atoms in the molecule of benzene; we might then expect phenol to be an alcohol. Experiments shew that some of the reactions of phenol are the reactions of an alcohol, but that other reactions are those of an acid. The compositions of methylic alcohol, ethylic alcohol, and phenol are all expressed by the formula R.OH ; in the first compound $R = CH_3$, in the second $R = C_2H_5$, and in the third $R = C_6H_5$: the fact that the third compound is an acid, as well as an alcohol in some reactions, must be connected with the composition of the group of atoms C_6H_5. We express our conception of the structures of the molecules $CH_3.OH$ and $C_2H_5.OH$ by the formulæ

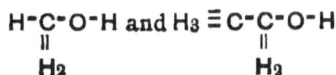

The arrangement of the atoms of carbon and hydrogen (leaving out of sight the group OH) in the molecules is thought of as very different from that of the atoms of carbon and hydrogen in the molecule $C_6H_5.OH$. That the difference between the properties of phenol and those of ethylic alcohol is to be connected with the arrangement of the atoms of carbon and hydrogen, other than those of the group OH, rather than with the numbers of these atoms, is made clear by the fact that the sixth hydrocarbon of the paraffin series, C_6H_{14}, gives a derivative $C_6H_{13}.OH$ which is a true alcohol and shews no acidic reactions. In phenol we have another example of the proposition that the functions of this or that atom in a molecule are dependent, among other conditions, on the arrangement of all the atoms which form the molecule.

It is because of its distinctly acidic character that phenol is so often called carbolic acid.

The production of nitrobenzene ($C_6H_5.NO_2$) from benzene, and of aniline ($C_6H_5NH_2$) from nitrobenzene, will be glanced at in the next chapter.

An examination of the formulæ given to benzene and salicylic acid shews that to pass from the first of these compounds to the second it is necessary to remove two atoms of hydrogen from the molecule of benzene and to substitute therefor the atomic groups OH and CO.OH. This process is effected thus: in the first place phenol is made from benzene; then sodium phenylate ($C_6H_5.ONa$) is prepared; then this substance (which is a solid) is heated in a stream of carbon dioxide to a temperature of about 130° C. [= 266° F.], when the following reaction occurs : $C_6H_5.ONa + CO_2 = C_6H_4.OH.CO_2Na.$; lastly the sodium salicylate thus obtained is decomposed by the proper quantity of a solution of hydrochloric acid, whereby a solution of sodium chloride and crystals of salicylic acid are produced. This is the method by which salicylic acid is manufactured. A salt of salicylic acid is contained in oil of winter-green, and what is called 'natural salicylic acid' is obtained from this salt.

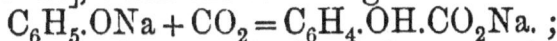

From time to time differences have been noticed in the actions of natural and artificial salicylic acid when administered medicinally. There ought to be no differences ; for the compositions of the two things are identical. Now if the reader will consider the following structural formulæ he will see that the hexagon formula for benzene provides for the existence of three isomeric acids of the composition $C_6H_4.OH.CO_2H$:—

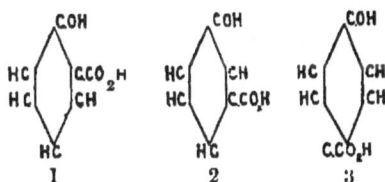

These acids must have different reactions, and almost certainly different therapeutic effects. The reactions of both natural and artificial salicylic acid shew that the first formula is the formula for that acid. Careful experiments have shewn that when sodium phenylate is heated in carbon dioxide to about 200° C. [= about 390° F.] a certain amount of the acid represented by the third formula is produced. Hence unless the temperature is carefully regulated in the preparation of salicylic acid, the product is a mixture of this acid with one of its isomerides. Determination of the percentage of each element in the product could not detect the presence of the isomeric acid ; and therefore it would be easy to pass as pure salicylic acid a substance which was really a mixture of that acid with another compound.

In this chapter we have again had examples of the very great influence which is exerted on the functions performed by carbon in the compounds whereof it forms a part by the relations between it and the other elements wherewith it may be united. There is no element which is so much affected by its companions as carbon ; in all the wanderings of the atoms of carbon, it matters much what these companions are, how many of this or that kind there are, and whether they crowd round the carbon atoms or stand apart from them, whether they surround the carbon atoms in solid phalanxes or draw themselves out in scattered chains.

CHAPTER XI.

THE chief changes of composition that occur in
the preparation of beer, of whiskey, and of wines
are changes that begin with sugars and end with
alcohols and ethereal salts. The main stages in beer-making are, the con-
version of barley into malt, the extraction by
water of the soluble constituents of malt, fermentation, and clearing. Barley is moistened and
allowed to germinate slightly; during this process a substance is formed, called *diastase*, which,
acting in the presence of water, converts the
starch of the barley into *dextrin*, and that variety
of glucose called *maltose*, and also slightly alters
the compositions of a group of nitrogen-containing compounds in the barley so that the
products dissolve in water. The composition of
dextrin is the same as that of starch $(C_6H_{10}O_5)$;
but, unlike starch, dextrin easily dissolves in
water. As maltose belongs to the group of the
glucoses, its composition is expressed by the
formula $C_6H_{12}O_6$. The germination is stopped
by heating the barley in a kiln; the product is
called *malt*. The malt is crushed, and treated
with water at about 60° to 70° C. [= 140° to
158° F.]; the liquor obtained by this process of
mashing is called *wort*. The wort contains
glucose, dextrin, soluble nitrogenous compounds,

153

and some mineral salts. Experience shows that
the changes from starch and insoluble nitrogenous compounds to dextrin, glucose, and
soluble nitrogen-containing bodies, proceed most
favourably at the temperature of 64° to 67° C.
[= 147° to 152° F.]. The wort is then boiled
with hops; certain substances which give a
peculiar flavour to the liquid are extracted from
the hops by the boiling wort. The liquid is
now rapidly cooled; if this is not done changes
occur in the constituents of the wort, and various
acids are formed that render the beer undrinkable. During the cooling, air finds its way into
the wort, and the oxygen in this air aids the
process of fermentation to which the wort is now
subjected. The fermentation—that is, the con-
version of glucose into alcohol and carbonic acid
gas—is accomplished by adding yeast to the wort,
and maintaining a suitable temperature. In
English breweries fermentation is conducted at
about 15·5° to 21° C. [= 60° to 70° F.]; in
German breweries, at about 12° to 15° C., or
sometimes at temperatures as low as 6° to 8° C.
[53° to 59° F., or 43° to 46·5° F.]. When fermentation occurs at 60° to 70° F., carbonic acid
gas escapes rapidly from the liquor, and particles
of yeast are carried towards the top of the fermenting wort; when the process is conducted at
43° to 53° F. the carbonic acid gas forms slowly,
and the yeast sinks to the bottom of the liquor.
The two processes are called *high fermentation*
and *low fermentation*, respectively. As has been
said, high fermentation is practised in England,
and low fermentation in Germany. The fer-

mented liquor is run off into cleansing tanks, where various insoluble substances gradually settle, and the yeast is skimmed off from the surface ; after a second process of settling and cleansing, the beer is run into barrels, and these are sent into the market.

The following figures present the quantities of the chief constituents in different kinds of beer :—

	Burton Ale.	Scotch Ale.	Lager Beer.	Porter.	Bavarian draught-beer
Per cent. of Water	79·6	81·5	90·8	86·3	90·26
,, Alcohol	5·9	8·5	3·7	6·9	3·8
,, Carbonic acid	..	0·15	0·22	0·16	0·14
,, Sugar* ,, Dextrin ,, Ash ,, Nitrogenous compounds	14·5	9·85	5·28	6·64	5·8

* The quantity of sugar in beer rarely amounts to more than a half per cent. ; porter may contain from 1·5 to 2 per cent.

Unless beer is made with great care the finished product is subject to various diseases. These diseases consist in the production of compounds that give a disagreeable taste and odour to the beer. The chief causes of these obnoxious changes are impurities in the yeast employed for effecting the fermentation. *Yeast* is the name given to a group of very simple plants of a low order ; these plants grow in wort, feeding on some of the constituents of the wort, and producing, during their growth, alcohol, carbonic acid, and traces of other compounds. Among the constituents of ordinary yeast are certain minute funguses, collectively known as *wild yeast,* which change alcohol into acetic acid, and glucose into lactic acid and butyric acid, and effect other transmutations, the pro-

ducts of which are more or less objectionable to
the beer-drinker. The cure for the disease of
beer is, therefore, the very careful selection of
yeast free from those wild varieties which bring
about the undesirable chemical changes. Un-
fortunately, it is impossible to detect wild yeast
in samples of the fungus microscopically; but if
the samples are grown under very definite conditions, and a microscopic examination is made
periodically of the growing organisms, it is
possible to distinguish the wild yeasts from those
which produce only alcoholic fermentation, for
the manner of growth is different in the different
kinds of yeast.

Whiskey is, or ought to be, made from malt.
The object of the distiller is to convert the whole
of the starch of the barley into glucose, and the
whole of the glucose into alcohol. The wort is
subjected to a brisk fermentation under conditions which ensure the transformation of the
whole of the sugar into alcohol; and sugar is
sometimes added to the wort, so that the fermented liquor may be richer in alcohol than it
would be if only what may be called the natural
sugar of the wort were acted on by the yeast.
The fermentation of wort for making whiskey
should be conducted at as low a temperature as
possible, else *fusel oil* is produced. This name is
given to a group of alcohols, the chief of which is
amylic alcohol (the fifth alcohol of the ethylic
series $= C_5H_{11}.OH$); these alcohols are very intoxi-
cating and distinctly poisonous. As whiskey
matures the fusel oil (some of which is present in
all newly-made whiskey) slowly reacts with the

small quantities of acids that fermented liquors always contain, and forms ethereal salts (compare Chap. VI., p. 93); and the *bouquet* of well-matured whiskey is chiefly caused by these ethereal salts. The fermented wort is distilled, and the distillate is distilled again; a liquid is thus obtained containing from 61 to 77 per cent. of alcohol; this is generally diluted so as to contain about 55 per cent. of alcohol, and bonded at this strength. Irish whiskey is usually bonded at about 64 per cent. alcoholic strength.

The strength of whiskey, brandy, rum, and other spirits, is very generally stated as so much 'overproof,' or so much 'underproof'; for instance, rum is generally imported at the strength of *twenty overproof.* In former times the strength of spirit, that is, the quantity of alcohol in the spirit, used to be roughly tested by pouring the spirit on to gunpowder, and igniting the vapour; if the burning vapour set fire to the powder the spirit was said to be *overproof*; if the powder remained unfired the spirit was said to be *underproof.* Hence arose the name *proof spirit*; spirit of such alcoholic strength that it would just fire gunpowder when it was poured on the powder and the vapour was ignited. As more accurate methods of estimating alcohol in liquids were worked out, it was found that proof spirit contained approximately 50 per cent. of alcohol and 50 per cent. of water. *Proof spirit* is now defined as a mixture of 49·24 parts by weight of alcohol with 50·76 parts by weight of water. A sample of whiskey containing 60 per cent. of alcohol, by weight, contains approximately 10 per

cent. more alcohol than proof spirit; as one part
of alcohol is approximately equal to two parts of
proof spirit, that whiskey would be described as
twenty overproof.

Methylated spirit consists of nine parts rectified
spirits of wine (containing about 84 or 85 per
cent. alcohol) mixed with one part methylic
alcohol (for a brief account of methylic alcohol,
or wood spirit, see Chapter V., p. 80). The
Inland Revenue authorities have allowed this
mixture to be bought and sold duty free, under
certain definite restrictions, since 1855. The
small quantity of methylic alcohol interferes very
slightly with the purposes for which methylated
spirit is used (for dissolving resins, for burning
in lamps, and in certain manufactures), it would
be extremely difficult to separate the ethylic
alcohol from the wood spirit in the mixture, and
the presence of the wood spirit is supposed to
make the liquid undrinkable. Of late years the
Board of Inland Revenue have added a small
quantity of oil to methylated spirit; if the spirit
is poured into water it forms a milky liquid with
a nasty smell.

Wine is the fermented juice of grapes. The
fermentation is generally conducted in open vats,
in summer, at about 10° to 12° or 14° C. [= 50°
to 57° F.], and the process occupies from 10
to 14 days. The air always contains sufficient
yeast-cells to bring about the change of the
sugar of the juice into alcohol and carbonic
acid.

The carbonic acid escapes; the liquid is racked
off into casks where it stands for some time, and

it is then transferred to other casks where it matures. Many chemical changes occur as wine matures ; there is a slow transformation of sugar into alcohol, various nitrogenous compounds and certain salts (especially tartrate of potassium) are precipitated, traces of acids are formed and react with the alcohols present (for fermented grape-juice always contains several alcohols besides ordinary, or ethylic, alcohol) to produce ethereal salts to the presence of which the *bouquet* of wine is chiefly due ; and a great many other changes of composition, which have not been fully eluci-dated, take place.

Wine is a very complicated liquid ; it contains ethylic alcohol and traces of other alcohols, one of which is glycerin ; acids, especially malic, tannic, tartaric acids, and succinic acid ; very small quantities of different ethereal salts ; nitro-genous compounds ; colouring matters ; and in-organic salts, especially phosphates and chlorides of potash, soda, lime, and magnesia. The quantity of alcohol in natural wines (that is, wines to which alcohol has not been added after fermentation, nor sugar beyond that contained in the grape-juice before fermentation) varies from about 6 to about 12 per cent.; the amount of acid is from 0·3 to 0·7 per cent.; the ethereal salts amount to a few hundredths of a per cent. of the wine ; most of the natural wines of France and the Rhine are practically free from sugar, sherry contains about 2˙ per cent., port about 4 to 5 per cent., and in such sweet wines as Tokay the sugar may amount to 25 per cent.; there may be about $\frac{2}{10}$ths of a per cent. of mineral salts ;

and from 2 to 3 per cent. of *extractive matter*, including nitrogenous compounds.

Champagne is bottled before the fermentation is finished ; the carbonic acid which is produced as fermentation proceeds in the bottled wine, cannot escape, and dissolves in the liquid ; when the cork is withdrawn the carbonic acid passes into the air and makes the wine effervesce (compare the remarks on soda water in Chapter IV. p. 51). A great deal of liquid is manufactured to be sold as *champagne*, by adding sugar to inferior white wine, and pumping in carbonic acid gas under pressure.

We must now briefly consider the changes that are turned to account in the manufacture of various explosives from cellulose. As we learned in the last chapter, cellulose is the main constituent of all vegetable tissues, and it has the same composition as starch. The simplest formula that expresses the composition of this compound is $C_6H_{10}O_5$, but it is almost certain that the molecule of cellulose contains a fairly large multiple of six atoms of carbon, ten atoms of hydrogen, and five atoms of oxygen.

If cotton wool, which is approximately pure cellulose, is steeped in a mixture of concentrated nitric and sulphuric acids, a change occurs which may be expressed approximately by the following equation :—

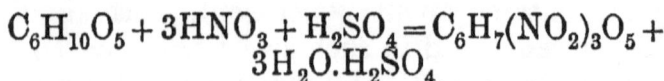

$$C_6H_{10}O_5 + 3HNO_3 + H_2SO_4 = C_6H_7(NO_2)_3O_5 + 3H_2O.H_2SO_4$$

Three atoms of hydrogen in the cellulose mole-

cule are replaced by three atomic groups, each
of which consists of an atom of nitrogen joined
to a couple of atoms of oxygen; the hydrogen
taken out of the cellulose molecule combines with
the rest of the oxygen of the nitric acid, and
the water that is so produced is seized by the
sulphuric acid wherewith it combines. The
nitrated product of this reaction $[C_6H_7(NO_2)_3O_5]$
is called *trinitrocellulose*; for all technical purposes it is known as *gun-cotton.*

When gun-cotton is dried and ignited in the
air it burns rapidly and is entirely changed into
gases; when it is fired in a closed space, although
it cannot obtain oxygen from the air, it contains
in itself sufficient oxygen to convert the whole of
the carbon in it to carbon monoxide and carbon
dioxide and a considerable quantity of the
hydrogen to steam. The other products of combustion under these conditions are hydrogen and
nitrogen; so that even in the absence of oxygen
outside itself, gun-cotton is wholly changed to
gaseous bodies when it is burnt, and the process
of burning takes place rapidly. The volume of
the gases produced by exploding gun-cotton,
after allowing the steam to condense, is about
750 times the volume occupied by the material
before the explosion. If gun-cotton is very
strongly compressed while wet a compact solid
is produced containing about 15 per cent. of
water : this substance is quite uninflammable,
and may be handled and moved about with perfect safety. Dry gun-cotton is easily exploded
by firing a small quantity of a detonator in
contact with it, such as fulminate of mercury

L

contained in a thin iron metal case, and the explosion spreads through the mass of gun-cotton with enormous rapidity. Moist, compressed, gun-cotton containing 15 per cent. of water cannot be exploded by a detonator unless a very large charge is employed; but if a small quantity of dry gun-cotton is brought into contact with the moist sample, and the dry substance is exploded by a detonator, the explosion spreads through the whole mass of the moist gun-cotton which is thereby completely and rapidly changed to gaseous substances.

It is evident then that gun-cotton possesses many of the properties of an ideal explosive; it is not very difficult to manufacture; under conditions which are easily attained it may be handled, carried, and stored, with complete safety; it can be exploded easily; the explosion spreads rapidly throughout the whole of the material, and produces an enormous volume of gaseous substances without the smallest quantity of any solid matter. Gun-cotton is the principal explosive used in the army and navy for effecting demolitions on land, for submarine mines, and for torpedoes.

It may be well to mention here another explosive, *nitroglycerin*, which is allied to gun-cotton chemically. This compound is made by treating glycerin $[C_3H_5(OH)_3]$ with concentrated nitric and sulphuric acids; the change may be expressed, approximately, thus:

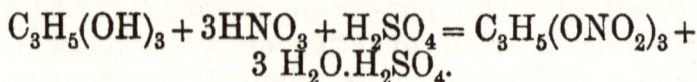

$$C_3H_5(OH)_3 + 3HNO_3 + H_2SO_4 = C_3H_5(ONO_2)_3 + 3\ H_2O.H_2SO_4.$$

Nitroglycerin is a highly explosive oil, dangerous to handle and to carry. *Dynamite* consists of a very fine naturally occurring silica soaked in nitroglycerin; the silica dilutes the nitroglycerin, so that the mixture can be safely handled and transported, and the solid dynamite is more convenient than the liquid nitroglycerin. Other diluents of nitroglycerin are employed, and most of the smokeless powders that are used at present are mixtures of nitroglycerin, or nitroglycerin and gun-cotton, with diluting materials which render the substances safe and convenient to handle, and also moderate the velocity of the explosion that occurs when the powder is fired by a detonator.

Let us now turn for a moment to some of those changes of composition and properties occurring among derivatives of benzene which are the bases of chemical industries. I will ask the reader's attention more particularly to the manufacture of aniline and aniline-colours.

Aniline, $C_6H_5.NH_2$, is made from nitrobenzene; and nitrobenzene, $C_6H_5.NO_2$, is made from benzene, which is one of the substances present in coal-tar. Hence the name *coal-tar colours* often applied to the colouring matters derived from aniline. Benzene is treated with a mixture of concentrated nitric and sulphuric acids, and the oily nitrobenzene so produced is thoroughly washed with water, and run into a still containing iron (in the form of scrapings from soft castings), and either acetic or hydrochloric acid, steam being blown into the still as the nitrobenzene runs in. A reaction begins at once; water, aniline, and unchanged nitrobenzene distil over ; more iron is added to the contents of the still, and after some time the distillate consists of only aniline and water. The aniline, which slowly separates from the water, is drawn off and redistilled. The principal reactions which occur in the manufacture of aniline from benzene may be expressed thus :—

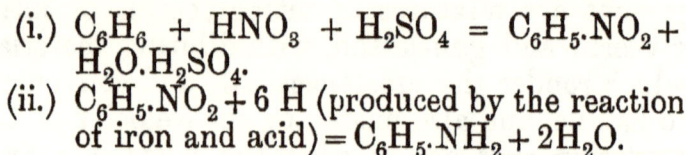

(i.) $C_6H_6 + HNO_3 + H_2SO_4 = C_6H_5.NO_2 + H_2O.H_2SO_4.$

(ii.) $C_6H_5.NO_2 + 6H$ (produced by the reaction of iron and acid) $= C_6H_5.NH_2 + 2H_2O.$

From aniline, which is a colourless oil, has been produced a vast number of derivatives and related compounds that are brilliantly coloured. For instance, the oxidation of commercial aniline, containing the compound $C_7H_7.NH_2$ (called *toluidine*), produces magenta-red or fuchsine. The constitutions of most of the aniline colours are very complicated. Some of them are substituted anilines, that is, compounds derived from $C_6H_5.NH_2$, by removing atoms of hydrogen and putting different more or less complex atomic groups in their place ; for instance, a yellow colouring matter is produced by treating aniline with the gas formed by warming nitric acid with starch ; this yellow body is derived from aniline by removing an atom of hydrogen from the molecule, $C_6H_5.NH_2$, and putting in its place the group of atoms $C_6H_5.N_2$. This yellow substance then becomes the starting point from which a series of coloured compounds is derived.

It is impossible to go into the constitution of
the aniline colours without possessing a very
minute acquaintance with the facts of organic
chemistry ; it must suffice to say here that the
whole of the aniline-colour industry is a direct out-
come of conceptions regarding the arrangements of
atoms in molecules, and the influences exerted
by the compositions and the relative positions of
various atomic groups on the properties of the
molecules into which they enter.

CHAPTER XII.

ALIZARIN AND INDIGO.

WE are told that, in times of war and when certain
religious ceremonies were being performed, our
ancestors used to dye their skins blue with the
juice of a plant called *woad ;* to-day we manufacture the blue dye-stuff of the woad in our
laboratories. For centuries large tracts of country
in the more southern parts of Europe were de-
voted to the growth of the plant whose roots were
used, under the name *madder*, for dyeing cloth
various shades of red : I do not suppose that there
is to-day more than a few acres of land in Europe
where this plant is cultivated, for one laboratory
now produces more of the dyeing compound of
madder, in a month, than was obtained in a year
from the plants grown on a thousand acres.
    The compound which gives its dyeing pro-
perties to woad is the same as that we call

indigo; the compound contained in madder is named *alizarin.* The name *indigo* is applied by Pliny to a pigment which came from India. The name alizarin is formed from the word *alizari,* the commercial name of madder in the Levant; this word is said to be derived from the Arabic *azara,* to press or squeeze, so that *alizari* would mean the pressed extract.

The composition of alizarin is expressed by the formula $C_{14}H_8O_4$. When the vapour of this compound is passed over heated zinc dust a hydrocarbon is obtained, which has the composition $C_{14}H_{10}$, and is called *anthracene.* This hydrocarbon is one of the many constituents of coal-tar; as coal-tar is produced in enormous quantities as a bye-product in making gas, there is here an almost unlimited source of alizarin, could a workable method be found for reversing the change that occurs by heating alizarin with zinc dust, and so producing alizarin from anthracene.

It was known, in the early sixties, that certain aromatic hydrocarbons were converted into compounds called *quinones* by the action of such oxidising agents as a mixture of sulphuric acid and bichromate of potash; in every case, the molecule of the quinone contained two atoms of oxygen in place of two atoms of hydrogen removed from the molecule of the hydrocarbon. It was also known that, in many cases, the hydrocarbon from which the quinone was obtained was produced by heating the quinone with zinc dust. In 1862 a compound was obtained by oxidising the hydrocarbon anthracene; the composition of anthracene is $C_{14}H_{10}$, and the composition of this product of its oxidation is $C_{14}H_8O_2$. The two chemists Graebe and Liebermann were busy in these days studying the relations of alizarin to other compounds. They made the guess that the compound $C_{14}H_8O_2$ was the quinone of anthracene; and that alizarin ($C_{14}H_8O_4$) was a derivative of *anthraquinone* related to that compound in the manner expressed by the formulæ

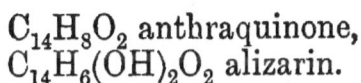

$C_{14}H_8O_2$ anthraquinone,
$C_{14}H_6(OH)_2O_2$ alizarin.

The facts then were these : anthracene, $C_{14}H_{10}$, could be oxidised without much difficulty to the compound $C_{14}H_8O_2$; alizarin had the composition $C_{14}H_8O_4$, and could be converted into anthracene by heating with zinc dust; several aromatic hydrocarbons could be oxidised to compounds containing the same number of atoms of carbon as the parent hydrocarbon, but two atoms of hydrogen less than that compound, and containing also a couple of atoms of oxygen; these products of oxidation of aromatic hydrocarbons were named quinones. Graebe and Liebermann assumed the compound $C_{14}H_8O_2$ to be the quinone of anthracene, and alizarin to be the dihydroxyl derivative of this quinone (the atomic group OH is call *hydroxyl*). One could begin with anthracene and make what was probably its quinone ; one could start from alizarin and arrive at anthracene ; there was one step untaken, that from the supposed quinone to alizarin. Experiments on the reactions of alizarin shewed that two of the eight atoms of hydrogen in the molecule of this compound were closely held to two of the four atoms of oxygen, and that the remaining six hydrogen atoms, and the remaining two oxygen atoms, were not directly joined to one another. (This is of course a statement of the results of experiments in the extremely symbolic language of the hypothesis of atom-linking.) In other words, one part of Graebe and Liebermann's hypothesis was justified. But how were two atoms of hydrogen to be removed from the molecule $C_{14}H_8O_2$ and two atomic groups OH to be put in their place? Many aromatic compounds were known to exchange atoms of hydrogen for an equal number of atoms of bromine when treated with bromine; and it was also known that such bromo-derivatives frequently exchanged their bromine atoms for an equal number of OH groups when fused with caustic potash. Graebe and Liebermann, therefore, treated the supposed anthraquinone ($C_{14}H_8O_2$) with the quantity of bromine calculated for the equation

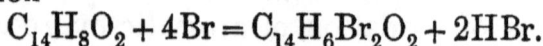

$$C_{14}H_8O_2 + 4Br = C_{14}H_6Br_2O_2 + 2HBr.$$

The product had the composition $C_{14}H_6Br_2O_2$. They then fused this compound with caustic potash, expecting the reaction

$$C_{14}H_6Br_2O_2 + 2KOH = C_{14}H_6(OH)_2O_2 + 2KBr$$

to occur. The product had the composition $C_{14}H_8O_4$, and the properties of alizarin.

The preparation of alizarin from anthracene was completed, and alizarin was proved to be *dihydroxyl-anthraquinone.*

But this method of making alizarin could not

be commercially successful, because bromine is very expensive, and very troublesome to manipulate. About this time (1866-68) Perkin, and independently of him Graebe and Leibermann also, applied to anthraquinone another reaction whereby atoms of hydrogen can be removed from the molecules of many aromatic compounds and their place taken by the atomic group OH. When benzene is heated with concentrated sulphuric acid a compound is produced which has the composition $C_6H_5.HSO_3$ and is called *benzene sulphonic acid.*

$$(C_6H_6 + H_2SO_4 = C_6H_5.HSO_3 + H_2O) ;$$

and when this acid is fused with caustic potash, phenol $(C_6H_5.OH)$ is formed.

Anthraquinone was heated with concentrated sulphuric acid for a little time, but no reaction occurred; very concentrated acid was used, and the heating was continued for a long time; at last a reaction began, and a compound was obtained whose composition and reactions shewed it to be *anthraquinone disulphonic acid.* The change may be expressed thus :—

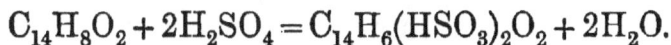

$$C_{14}H_8O_2 + 2H_2SO_4 = C_{14}H_6(HSO_3)_2O_2 + 2H_2O.$$

This acid (which was a solid body) was then fused with caustic potash; the product was alizarin :—

$$C_{14}H_6(HSO_3)_2O_2 + 2KOH = C_{14}H_6(OH)_2 + 2KHSO_3.$$

The results of many investigations into the re-

actions and relations of alizarin find expression
in the structural formula

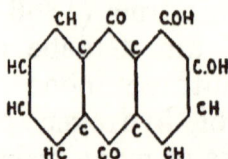

The hydrogen and oxygen atoms, and the two
atomic groups OH, are thought of as attached to
three benzene nuclei, with two carbon atoms
belonging to both the first and the second
nucleus, and two common to the second and the
third. But it is not possible to analyse this
formula and elucidate its meaning without a
more thorough knowledge of organic chemistry
than is to be looked for on the part of readers of
such a book as this. There are nine possible
isomerides of alizarin; but alizarin is the only
one which can be used as a colouring material.
It is a somewhat remarkable coincidence that the
result of the first attempt to prepare dihydroxyl
anthraquinone artificially should have been the
formation of that one isomeride which was of any
value as a dye-stuff.

Thirty years ago, about half a million tons of
madder were sent into the market annually, and
about half of that was grown in France. The
amount of madder exported from France ten
years ago was a few hundred tons. The discovery of a cheap method for making alizarin,
taken with the discovery of the aniline colours,
entirely revolutionised the trade in dyeing
materials. In his report on the Exhibition of
1862, Hofmann wrote:—

" England will, beyond question, at no distant day, become herself the greatest colour-producing country in the world ; nay, by the strangest of revolutions, she may, ere long, send her coal-derived blues to indigo-growing India, her tar-distilled crimsons to cochineal-producing Mexico, and her fossil substitutes for quercitron and safflower to China, Japan, and the other countries whence these articles are now derived."

Hofmann was both right and wrong. Coal-derived blues and tar-distilled colours of every shade are sent all over the world ; but they are not sent by England. Germany is the great colour-producing country of the world to-day. The artificial production of dyeing compounds is a triumph of the systematic study of the order which reigns in the domain of the almost infinitely minute ; but regarded from the artistic standpoint that triumph has been a calamity to mankind.

Indigo, a substance that has been used as a pigment, and also for dyeing, since very early times, is obtained from many plants most of which flourish in tropical regions. Indigo is imported chiefly from India, where it is manufactured from the plants of *indigofera tinctoria*, a perennial that is cultivated as an annual. The plants are steeped in water in large vats ; fermentation proceeds for several hours ; the yellow liquid is run off into other vats wherein it is agitated by paddles until the colour changes to dark blue and a blue sediment is deposited ; this sediment is thoroughly washed with boiling water, drained on canvas filters, pressed in shallow wooden frames, and cut into cubes which are dried in open-air sheds. The compound which gives its colour to indigo can be separated by a series of operations; it is a dark-blue crystalline powder with a bronzy lustre; the compound is called *indigotin*, and its composition and molecular weight are expressed by the formula $C_{16}H_{10}N_2O_2$.

When indigotin is subjected to the action of reagents that remove oxygen from compounds, an alkali also being present, it combines with two atoms of hydrogen and forms an almost white compound called *indigo-white*, $C_{16}H_{12}N_2O_2$, which dissolves in the alkali that is present. On exposing this yellowish liquid to the air, oxygen is slowly absorbed and the blue compound, indigotin, is reproduced as a solid. It is on these reactions that the dyeing of silk, wool, and cotton, with indigo is based. An *indigo-vat* is prepared by mixing indigo with water, an alkali (commonly slaked lime), and some deoxidiser such as green vitriol (sulphate of iron), or zinc powder, or hyposulphite of soda. The indigotin is reduced to indigo-white and this goes into solution; the goods are steeped in the nearly colourless liquid, and then exposed to the air; indigotin is slowly formed, by the action of atmospheric oxygen, and being deposited in the fibres of the cloth it remains firmly attached thereto. Indigo-vats are sometimes prepared by mixing the indigotin with water, lime, and materials which bring about fermentation; the result of the fermentation is indigo-white.

Indigotin has been made artificially by several processes, each of which involves a complex series of reactions. All that may be done here is to give the merest sketch of the oldest of these transformations. The study of the relations of indigotin to other compounds was undertaken by Baeyer about the year 1865, and the synthesis of the compound was effected by him after experiments which lasted more or less continuously for thirteen years. First it was shewn that indigotin $(C_{16}H_{10}N_2O_2)$ is oxidised to *isatin*, $C_8H_5NO_2$; then, in attempting to re-convert isatin into indigotin, various compounds were discovered intermediate between these two, more especially *oxindol*, C_8H_7NO; then oxindol was changed into isatin; and then isatin was deoxidised to indigotin $(2C_8H_5NO_2 + 4H = C_{16}H_{10}N_2O_2 + 2H_2O)$; finally oxindol was prepared from an acid (called *amidophenylacetic acid*), which itself was made from phenol, a compound that is found in large quantities in coal-tar. It was only necessary then to separate phenol from coal-tar, from the phenol to prepare the acid, from this to make oxindol, and then to convert oxindol into isatin, and isatin into indigotin.

Many other methods for making indigotin have been worked out, some of them considerably simpler than the original method of Baeyer. Several methods have been patented, but none has yet been a commercial success, because natural indigo is still cheaper than the artificially made substance.

CHAPTER XIII.

THE ALKALOIDS AND ALBUMIN.

A GREAT many compounds capable of forming salts by combining with acids, and having marked toxicological effects, have been obtained from plants. These compounds are classed together under the name *alkaloids,* because they resemble alkalis in some of their reactions. All the alkaloids except three are compounds of carbon with nitrogen, hydrogen, and oxygen; the three exceptions are *coniine* (the alkaloid of hemlock), *nicotine* (the alkaloid of tobacco), and *spartëine* (the alkaloid of the common broom); these three alkaloids are composed of carbon, nitrogen, and hydrogen only. Among the commoner alkaloids may be named *quinine* and *cinchonine,* obtained (with more than twenty other alkaloids) from Peruvian bark; *theobromine,* from cocoa; *caffeine,* from coffee; *thëine,* from tea; *nicotine,* from tobacco; *morphine* and *narcotine,* obtained (with about fifteen other alkaloids) from opium; and *strychnine* and *brucine* from *nux vomica.*

The formulæ which express the compositions of the alkaloids are generally rather complex; such as, $C_{20}H_{24}N_2O_2$ for quinine, $C_{21}H_{22}N_2O_2$ for strychnine, and $C_{17}H_{19}NO_3$ for morphine. The study of the reactions of the alkaloids, and their relations to other compounds whose chemical properties have been expressed in constitutional formulæ, has not yet advanced very far; although a vast number of observations and experiments

174

has been made. A few of the alkaloids have been
built up from simpler materials ; for instance,
coniine, *atropine* (the alkaloid of deadly night-
shade), caffeine, and theobromine. In working
on the reactions of atropine, a compound was
discovered closely related to that alkaloid, and
possessing, more pronouncedly than atropine, the
property of dilating the pupil of the eye. The
alkaloid *cocaine* (prepared from coca leaves) is
used as a local anæsthetic in smaller surgical
operations, especially in operations on the eye :
it seems certain that this alkaloid will be synthesised very soon ; indeed a compound has been
obtained with exactly the same composition as
cocaine but without any physiological action ;
when some alterations have been effected in the
spatial arrangement of the atomic groups in the
molecule of this inactive isomeride, we shall have
artificially made cocaine in the hands of the
oculists. After that will come the synthesis of
quinine and other fever-allaying compounds. A
distinguished chemist has told us that when he
asked the purpose of the large buildings he saw
being erected in a German chemical works, some
years ago, he was told :—" These are our future
quinine works." Already at least two drugs that
are most beneficial in reducing the temperature
of the body in fever-cases have been manufactured
artificially ; these are *antipyrine* and *phenacetin.*
The molecule of antipyrine is built up step by
step from benzene and acetic acid ; the reactions
that occur in the process of synthesis, taken with
those which are noticed when antipyrine is decomposed into simpler compounds, lead to the

representation of the intramolecular arrangement
of the atoms in this compound by the formula

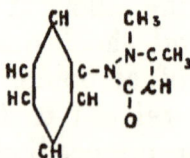

Antipyrine has all the chemical properties of an
alkaloid; it is an example of a body of this class
made in the laboratory, but not yet manufactured,
so far as we know, by a living plant.

The synthetical and analytical reactions of
phenacetin are expressed in the language of atom-
linking by the comparatively simple formula

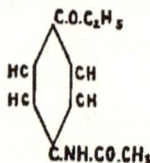

One of the hydrogen atoms of a benzene molecule is replaced by the atomic group ($O.C_2H_5$),
and another by the group ($NH.CO.CH_3$).

Although the constitution of quinine is not yet
fully worked out, it is certain that this alkaloid
is closely allied to two compounds named *quinoline*
and *pyridine* which are found in coal-tar. Quinoline has the composition C_9H_7N, and pyridine
the composition C_5H_5N. The reactions of pyridine find their expression in the constitutional
formula

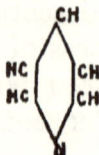

The reactions of quinoline are expressed by a
formula which represents the molecule of that
compound as composed of a benzene molecule
with two hydrogen atoms removed, and this
residue joined to a pyridine molecule from which
a pair of hydrogen atoms has been taken away;
thus

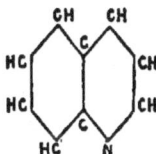

It is probable that the molecule of quinine
is formed by adding another pyridine ring to the
molecule of quinoline, and substituting various
atomic groups for some of the hydrogen atoms in
this arrangement. Such a formula as the following would express this conception of the structure
of the molecule of quinine.

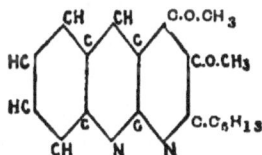

A general similarity may be detected in the
formulæ whereby we picture to ourselves the
connexions between the reactions and the molecular structures of the alkaloids that have been
studied most thoroughly. There seems to be a
foundation, either of a benzene nucleus, or a
benzene nucleus intimately linked to a similar
nitrogen - containing group of carbon and hydrogen atoms, and attached to this foundation
there are certain side chains composed in part of

the atomic groups that occur in the molecules of the ethylic alcohols and ethers (compare Chapter VI. p. 95).

Chemists are only feeling their way towards a knowledge of the relations of the alkaloids—those products of the activity of living plants—to other simpler bodies, towards a knowledge so exact as to be expressed in the clear, descriptive, and suggestive language of molecular structure. The method whereby that accurate knowledge is being gained is no new method : it consists in examining the reactions of this or that compound under definite conditions, and thereby finding other bodies to which the compound is related, both in that the compound is obtained from some of these other bodies, and some of them are obtained from the compound. The reactions of such a substance as quinine are exceedingly many ; but most of the recorded reactions seem to throw little light on the structure of the molecule of the substance. I have said that the method employed in trying to elucidate the structure of the molecules of such complicated bodies as quinine and other alkaloids is the method which has been so fruitful when applied to bodies of less molecular complexity ; nevertheless, it may be necessary to devise new ways of applying the method when one is dealing with compounds which are the balanced products of many most delicate transformations. As the problems of intramolecular arrangement become finer and more subtle the plan of attack must be marked by greater mobility and more finesse.

We have become acquainted with many facts

concerning composition and properties which can be reduced to order, and thought of clearly as related one to another, at present at any rate, only by stating them in terms of the molecular and atomic theory: no. mechanism has yet been devised whereby we can definitely connect changes in the properties, with changes in the compositions, of compounds, except that of the molecule and the atom. And we have more and more come to think of the molecule as a system in equilibrium because of the mutual actions and reactions between its parts ; as a system of atoms, and groups of atoms, wherein the function performed by one part is modified by the functions of all the other parts. The conception of molecular symmetry is one of the leading guides to-day to those who are trying to throw more light on the influence of composition on the properties of compounds, and especially compounds of carbon. A certain atomic group is introduced into a molecule of a carbon compound ; the product has marked dyeing powers ; it dyes a deep yellow: another atomic group is introduced in place of what may be called the yellow group of atoms ; the product dyes blue. It is wished to prepare a compound which shall dye purple ; this is effected by balancing the influence of the yellow group of atoms by such a number of blue groups as to produce a molecule which is a purple dye-stuff. The series of experiments is repeated but under somewhat different conditions, perhaps the temperature whereat the changes are effected is higher than it was in the first series of experiments ; the product has the

same composition as the purple-dyeing compound,
but it does not dye at all. Another carbon compound is known allied to that from which the
bodies that dyed yellow, blue, and purple were
obtained ; the yellow group of atoms is introduced into the molecule of this compound, and
the result is a body that dyes a pale, feeble, and
undesirable yellow. Evidently, then, it is not
accurate to speak of a certain group of atoms as
a yellow-dyeing group, and of another group as
a blue-dyeing group : whether tinctorial effects
do or do not accompany the presence of these
groups in molecules of carbon compounds, and
what the tinctorial effects are, evidently depend
on the positions of the groups in a molecule
relatively to other groups and atoms, and also on
what are the compositions of these other groups,
and on the nature of the other atoms. It is only
when the yellow group, or the blue group, is
introduced into a molecule of appropriate structure, and is placed in an appropriate position in
that molecule, that the new molecule dyes yellow
or blue : it is only when the yellow group is
balanced by the blue group, by reason of the
relative positions of these groups, and this
balancing is effected in a molecule wherein the
arrangement and also the compositions of the
other parts are such as to exert a certain influence on the functions of the two dyeing groups,
it is only then that the new molecule dyes
· purple. A vague general conception of molecular symmetry would be of no use in practical
science ; it is because this conception is transformed by the hypotheses of atom-linking and

definite atom-fixing power into a working instrument of research that it has become so important in advancing the knowledge of the interdependence of composition and properties.

In Chapter IX. I attempted to sketch, in outline, the application to the tartaric acids and certain sugars of the conception of the asymmetric carbon atom. That hypothesis, and the formulæ which arise from it, pictured the differences between the structures of the tartaric acids as the difference between an object and its reflection in a mirror. If some of the figures given in Chapter IX. are considered (see especially p. 139), the reader will see that a change in the structure, and hence in the properties, of such a compound as tartaric acid might be caused by the semi-rotation of one of the tetrahedra, that is, one of the groups of a carbon atom attached to four different atoms and atomic groups; for the partial rotation would alter the positions of the atoms and groups belonging to the rotated tetrahedron relatively to the atoms and groups of the other, unmoved tetrahedron. Changes in the structure, and the properties, of compounds that exhibit geometrical isomerism (see p. 135) might theoretically occur with very small changes of external conditions. Now experiments show that many geometrically isomeric compounds are very ready to undergo small changes which do not alter the percentage compositions or the molecular weights of the compounds. There is one especially interesting example of such changes. The existence of two isomerides of a certain compound was asserted by a German chemist of recognised accuracy and ability; the existence of one of these isomerides was denied by another chemist of equal capacity and equal accuracy. One of the chemists said he had obtained both forms of the compound; the other chemist replied that he could obtain one form only. If the two forms of the compound existed, it was necessary to think of them as geometrical isomerides, as one the mirror-image of the other. After a good deal of controversy each worker published the exact details of his experiments; one had worked on a bench exposed to full sunshine, the other in a part of the laboratory screened from the direct rays of the sun. The shade-loving chemist repeated the experiments in the sunshine, and obtained two modifications of the compound. The interpretation given by the hypothesis of geometrical isomerism is simple; under the influence of the direct sunshine a semi-rotation of parts of some of the molecules of the compound had occurred; when the sunshine was excluded all the molecules were geometrically identical.

This example shows how fine are some of the problems concerning the connexions of properties with structure, and how delicate the instruments must be whereby these problems are to be solved. And as we approach those bodies which seem to be very intimately associated with the maintenance of life in animals and plants, the problems become yet finer and more difficult of solution. *Albumin* is the name given to a class of compounds of carbon, hydrogen, oxygen, nitrogen, and sulphur which form a section of a larger class of compounds of these five elements named *proteids*. Proteids are always present in the protoplasm of living, active, cells. The percentage composition of different proteids varies within not very wide limits. Analyses of the same proteid, for instance, of egg albumin, show distinct differences ; this may be because the specimens analysed contained varying quantities of impurities, or because the substance we call albumin is a mixture of different compounds, or because the compositions of the proteids in a living organism are constantly changing. If the third hypothesis is adopted, then the proteids cannot be classed among chemical compounds, if the term *compound* is employed with its ordinary signification. But, one may say, if a body is not a compound, if it has not a definite, unchangeable composition, it is a mixture ; and if proteids are mixtures, the chemist is not concerned with them, at least not until they have been separated into their constituent compounds. In the study of natural occurrences it is well constantly to remind oneself that " in nature there are no boundary lines, however necessary it may be for us to draw them " ; and " in nature everything is distinct, yet nothing defined into absolute independent singleness." A change in the arrangement of a group of atoms so small that we can only liken the product of that change to the image of an object reflected in a mirror is accompanied by a very distinct change of properties. May we not refine a little more ? May we not picture to ourselves a complex collocation of atoms constantly giving up a few atoms to the substances which environ it, and constantly assimilating some of the atoms which compose the materials of its environment ? May we not speak of that complex atomic group as existing only as long as it undergoes change of this kind, as exhibiting its characteristic properties only while it is in a state of flux; as being, only when it is becoming; as existent, only as it is ceasing to exist; as finding itself only by losing itself ? The difficulty is to translate such loose conceptions as these into a working hypothesis for use in the laboratory. We cannot do this to-day, we may do it to-morrow.

Inorganic chemistry—the chemistry of the metals and of mineral compounds—presents phenomena not wholly unlike those set before us by the chemistry of the constituents of living protoplasm. The presence of a minute trace of phosphorus enormously modifies the properties of steel; and the same specimen of steel changes its properties very markedly when it has been used for some time. Sodium may be distilled in oxygen without a trace of a compound being formed, if both substances are perfectly dry; but if there be a very minute trace of water present, combination occurs violently. Electric sparks may be passed through a perfectly dry mixture of carbon monoxide and oxygen gases, and no change occurs; let an infinitesimal quantity of water be added, the carbon monoxide and oxygen disappear, and carbonic acid gas is formed in their place.

Should the student of natural science succeed in removing the boundary line that has been

SUMMARY AND CONCLUSION. 185

drawn between the phenomena of living matter and the phenomena of non-living matter, the removal will deepen our realisation of the unity of nature, and increase our feeling of delightful wonder.

CHAPTER XIV.

SUMMARY AND CONCLUSION.

BEFORE the facts about the compositions of compounds had been generalised in the laws of combination, the knowledge of the composition of a body gained by analysis was expressed by stating the quantity by weight of each element in one hundred parts of the compound analysed. These were the days of facts, and facts only. When a theory of the structure of matter had become a light to the feet of the investigator, the results of the analyses of compounds were expressed in general statements, and the laws of chemical combination brought order into the chaos of facts. The compositions of compounds were now expressed in terms of the number of combining weights of each element in that quantity of a compound which reacted with other compounds to produce new bodies. And then the introduction of the illuminating conceptions of the atom and the molecule enabled an exact meaning to be given to the indefinite expression *that quantity of a compound which reacts with other compounds to produce new bodies.*

A clear picture could now be formed of the mechanism of chemical changes : the chemist could see, mentally, the clashing of molecules, the disintegration of these minute particles, and the re-arrangement of the parts, the atoms, in new collocations which are new molecules. The atomic and molecular theory is the only guide that has been found equal to the task of leading the chemist through the maze of facts, and especially the facts concerning the compositions and properties of compounds of carbon.

I have asked the reader of this book to follow the wanderings of certain atoms and to pay heed to some very striking changes of properties which accompany the association of these minute bodies with other particles of like magnitude with themselves. I have tried to show that a vast number of facts about changes of composition and changes of properties are brought into due order and are related to one another by the hypothesis that the properties of those collocations of atoms we call molecules are conditioned by the nature, the number, and the arrangement, of the individual atoms, and that very great changes in the properties of molecules may be effected by very small alterations in the grouping of the atoms whereof the molecules are composed. I have sketched the treatment by which the conception of molecular structure has been made into a potent instrument for framing general expressions of the similarities and dissimilarities between reactions, for forming clear pictures of the ways wherein composition and properties are linked together, and for helping the memory to retain the results of the experimental investigations of chemical changes.

In applying the conception of molecular structure to express observed relations between properties and composition, the parts of molecules were represented, for a time, as arranged in two dimensions in space; and it was found possible for many years to present all the observed facts in terms of this admittedly imperfect hypothesis. But reactions and properties of compounds were observed which could not be translated into two-dimensional formulæ; and so chemists were obliged to refine their methods of presenting experimental results in structural formulæ. The consideration of the three-dimensional formulæ that were fashioned to present the finer differences between isomeric compounds suggested the possibility of changing one isomeride into another without effecting any disintegration of the molecule. A change so small that it could be likened to giving a half-turn to one of a pair of tetrahedra joined at one summit of each was found often to be sufficient to alter some of the physical properties of a compound. And as chemists are beginning to see into the conditions of existence of those chemical substances which are very intimately connected with living cells, it is evident that their conceptions of molecular structure must be refined much more.

At one time differences of properties were associated with the removal of one or more atoms from a molecule, and the wandering of these atoms into other molecules. Then the migration of an atom, or a group of atoms, from one position to another in the same molecule was recognised as carrying with it a change in the properties of the molecule. Then it was found necessary to admit that changes of properties may occur if a slight twist is given to one half of a molecule, the two halves of which are counterparts of one another. And already chemists see that their conceptions of molecular structure, and changes of properties as accompaniments of atomic wanderings, must be made much finer and more delicate if they are to embody the facts that are being accumulated in their laboratories. As, formerly, it was found possible to present many facts in terms of an hypothesis which was admittedly quite insufficient to explain the facts—the hypothesis, namely, that the parts of molecules are arranged in two dimensions in space ; so, now, even the most refined formulæ, whereby slight differences of molecular structure are presented in a consistent and suggestive manner, wilfully ignore a matter of fundamental importance. All structural formulæ represent the parts of molecules as fixed relatively to one another ; but we are sure that these parts are constantly in motion. We simplify knowingly, that the problem may be made amenable to accurate treatment. It is impossible to attack the problems of natural science otherwise than by simplifying them : we cannot give an exact and complete statement, much less a complete explanation, of any natural occurrence ; it is literally true that—

" I hold you here, root and all, in my hand,
Little flower—but *if* I could understand
What you are, root and all, and all in all,
I should know what God and man is."

We cannot yet devise formulæ which shall present the molecule as a system whose parts are constantly performing regulated movements; therefore, we make shift to do with formulæ which picture these parts as fixed relatively to one another. And this crude device works well on the whole ; for, whatever may be the motions of the parts of molecules, certain relations are always maintained between the parts ; and it is the mutual relations of the atoms, and the groups of atoms, that form the subject of chemical investigation.

The atoms are always wandering ; but as long as their excursions follow a certain sequence, and are confined with a certain limit, no change occurs in the properties of the molecule. One atom, or a group of atoms, may wander beyond its normal limits, and its relations to the rest of the molecule may become permanently altered. An atom, or an atomic group, may wander outside the molecule, and become part of another molecule. These are the kinds of the wanderings of atoms. To distinguish these wanderings, to express them in lucid and suggestive language, and to connect some of them with definite changes in the properties of molecules, is the business of chemical science.

INDEX.

www.ingramcontent.com/pod-product-compliance
Lightning Source LLC
Chambersburg PA
CBHW021711210326
41599CB00013B/1613